4TH GENERATION WARFARE HANDBOOK

D0916733

CASTALIA HOUSE

NON-FICTION

SJWs Always Lie: Taking Down the Thought Police by Vox Day
Cuckservative: How "Conservatives" Betrayed America
    by John Red Eagle and Vox Day
On the Existence of Gods by Dominic Saltarelli and Vox Day
Equality: The Impossible Quest by Martin van Creveld
A History of Strategy by Martin van Creveld
4th Generation Warfare Handbook
    by William S. Lind and LtCol Gregory A. Thiele, USMC
Four Generations of Modern War by William S. Lind
On War: The Collected Columns of William S. Lind 2003-2009
Transhuman and Subhuman by John C. Wright
Between Light and Shadow: The Fiction of Gene Wolfe, 1951 to 1986
    by Marc Aramini
Compost Everything by David the Good
Grow or Die by David the Good
Astronomy and Astrophysics by Dr. Sarah Salviander

MILITARY SCIENCE FICTION

There Will Be War Vol. X ed. Jerry Pournelle
There Will Be War Vol. IX ed. Jerry Pournelle
Riding the Red Horse Vol. 1 ed. Tom Kratman and Vox Day

SCIENCE FICTION

Somewhither by John C. Wright
Awake in the Night Land by John C. Wright
City Beyond Time: Tales of the Fall of Metachronopolis by John C. Wright
Back From the Dead by Rolf Nelson
QUANTUM MORTIS A Man Disrupted by Steve Rzasa and Vox Day
QUANTUM MORTIS A Mind Programmed
    by Jeff Sutton, Jean Sutton, and Vox Day
Victoria: A Novel of Fourth Generation War by Thomas Hobbes

FANTASY

Iron Chamber of Memory by John C. Wright
The Book of Feasts & Seasons by John C. Wright
A Throne of Bones by Vox Day
Summa Elvetica: A Casuistry of the Elvish Controversy by Vox Day
The Altar of Hate by Vox Day

AUDIOBOOKS

A History of Strategy, narrated by Jon Mollison
Cuckservative, narrated by Thomas Landon
Four Generations of Modern War, narrated by William S. Lind
Grow or Die, narrated by David the Good
Extreme Composting, narrated by David the Good
A Magic Broken, narrated by Nick Afka Thomas

4TH GENERATION
WARFARE HANDBOOK

WILLIAM S. LIND AND
LTCOL GREGORY A. THIELE, USMC

4th Generation Warfare Handbook

William S. Lind and Gregory A. Thiele

Published by Castalia House
Kouvola, Finland
www.castaliahouse.com

Cover Design: JartStar
Cover Image: Ørjan Ruttenborg Svendsen

ISBN: 978-952-7065-75-4

Contents

Foreword   1

Introduction   3

1   Understanding Fourth Generation War   7

2   Fighting Fourth Generation War   39

3   The Fourth Generation Warfare Counterforce   63

4   Light Infantry Tactics   69

5   Light Infantry Training Objectives   73

6   Operation Pandur   77

7   Training Light Infantry Units   93

8   Light Infantry Conversion Training Plan   101

9   Meeting the Fourth Generation War Challenge   109

Appendix A: The First Three Generations of Modern War   115

Appendix B: The 4GW Canon   119

Appendix C: Light Infantry Essential Reading   123

Notes   125

Foreword

This book is a follow-on to William S. Lind's *Maneuver Warfare Handbook*, first published in 1985 and still in print.

Its genesis lies in a series of seminars initially led by Mr. Lind and later co-chaired by several Marine officers, most frequently then-Major Greg Thiele. The first seminar, which met at Mr. Lind's home in Alexandria, Virginia, created the first Fourth Generation war field manual, *FMFM-1A, Fourth Generation War*. Subsequent papers dealing with specific aspects of Fourth Generation war were written by seminars at the Marine Corps' Expeditionary Warfare School in Quantico, Virginia. These studies were published as manuals of the Imperial and Royal (K.u.K.) Austro-Hungarian Marine Corps and are available at www.traditionalright.com.

The seminars were composed primarily of Marine Corps officers, mostly captains, plus officers from each of the other U. S. Armed Services and foreign officers from services including the Royal Swedish Marines, Royal Marines, Royal Netherlands Marines and the Argentine Army and Marine Corps. We gratefully acknowledge the work of all these officers and hope this handbook will serve to spread their thoughts on Fourth Generation war to a wider audience.

This book presumes a familiarity with the overall framework of the Four Generations of Modern War, a framework created by Mr. Lind in the 1980s. Those who wish to study the overall framework, including the first three generations, before reading this handbook can find a brief description included in the appendices. A more detailed description is available in both audiobook and ebook form, *The Four Generations of Modern War*, from Castalia House press.

William S. Lind
LtCol Gregory A. Thiele, USMC

Introduction

Just as Alexander's exploits only reached the Middle Ages as a dim, fantastic tale, so in the future people will probably look back upon the twentieth century as a period of mighty empires, vast armies and incredible fighting machines that have crumbled into dust....

—Martin van Creveld, *The Transformation of War*

War is Changing

War always changes. Our enemies learn and adapt, and we must do the same or lose. But today, war is changing faster and on a larger scale than at any time in the last 350 years. Not only are we facing rapid change in *how* war is fought, we are facing radical changes in *who* fights and what they are fighting *for*. All over the world, state militaries find themselves fighting non-state opponents.

This kind of war, which we call Fourth Generation war, or 4GW, is a very difficult challenge. Almost always, state militaries have vast superiority over their non-state opponents in what we call "combat power": technology, weapons, techniques, training, etc. Despite this superiority, more often than not, state militaries end up losing.

America's greatest military theorist, U. S. Air Force Colonel John Boyd, used to say, "When I was a young officer, I was taught that if you have air superiority, land superiority and sea superiority, you win. Well, in Vietnam we had air superiority, land superiority and sea superiority, but we lost. So I realized there is something more to it."

This handbook is about that "something more." In order to fight Fourth Generation war and win, the armed forces of the state need to understand what that "something more" is. That, in turn, requires an intellectual framework—a construct that helps us make sense of facts and events, both current and historical.

The intellectual framework put forward in this handbook is called "The Four Generations of Modern War."[1] This concept was first laid out in an article in the *Marine Corps Gazette* in October, 1989.[2] In this framework, modern war began with the Peace of Westphalia in 1648 which ended the Thirty Years War. Why? Because with that treaty the state—which was itself a relatively new institution[3]— established a monopoly on war. After 1648, first in Europe and then around the world, war was transformed into an activity that was waged by states against other states, using state armies, state navies, and eventually, state air forces. To modern observers, the assumption that war is something waged by states is so automatic that we have difficulty thinking of war in any other way. We sometimes misleadingly call war against non-state opponents "Operations Other Than War" or "Stability and Support Operations," or more recently, "Hybrid War."

In fact, before the Peace of Westphalia, many different entities waged wars. Families waged wars, as did clans and tribes. Ethnic groups and races waged war. Religions and cultures waged war. So did business enterprises, legal as well as illegal. These wars were often many-sided, not two-sided, and alliances constantly shifted.

Not only did many different entities wage war, they used many different means. Few of these non-state entities possessed anything we would recognize as a formal army, navy or Marine Corps, although Marines were often present in the form of fighting men on galleys. When war came, whoever was fighting would hire mercenaries both on land and at sea. In other cases, such as tribal war, the "army" was any male old enough to carry a weapon, but not too old to fight. In addition to campaigns and battles, war was waged by bribery, assassination, treachery, betrayal, and even dynastic marriage. The lines between civilian and military, and between crime and war, were either hazy or nonexistent. Many societies knew little internal order or peace; when not hired out for wars, bands of men with weapons simply took whatever they wanted from anyone too weak to resist them.

Here, the past is prologue. Much of what state-armed forces now face in Fourth Generation wars is simply war as it was fought before the rise of the state and the Peace of Westphalia. Once again, clans, tribes, ethnic groups, cultures, religions and gangs are fighting wars, in more and more parts of the world. They fight using many different means, not only conventional engagements and battles. Once again, conflicts have become many-sided rather than two-sided. Officers and enlisted men who find themselves caught up in such conflicts quickly discover they are difficult to understand and even harder to win.

The Root of the Problem

At the heart of this phenomenon, Fourth Generation war,[4] lies not a military evolution but a political, social, and moral revolution: a crisis of legitimacy of the state. All over the world, citizens of states are transferring their primary allegiance away from the state to other entities: to tribes, ethnic groups, religions, gangs, ideologies, and "causes."

Many people who will no longer fight for their state are willing to fight for their new primary loyalty. In America's two wars with Iraq, the Iraqi state-armed forces showed little fight. But Iraqi insurgents whose loyalties were to various non-state elements waged a hard-fought and effective guerrilla war.

The fact that the root of Fourth Generation war is a political, social, and moral phenomenon, the decline of the state, means there can be no purely military solution to Fourth Generation threats. Military force is incapable, by itself, of restoring legitimacy to a state.

This is especially the case when the military force is foreign; usually, its mere presence will further undermine the legitimacy of the state it is attempting to support. At the same time, there can be little doubt that state-armed forces will be tasked with fighting Fourth Generation wars. This is not just a problem, it is a dilemma—one of several dilemmas state militaries will face in the Fourth Generation of modern war.

With this dilemma constantly in view, the *4th Generation Warfare Handbook* explains how to prepare for and fight Fourth Generation war.

Chapter 1

Understanding Fourth Generation War

The first, the supreme, the most far-reaching act of judgment that the statesman and commander have to make is to establish … the kind of war on which they are embarking; neither mistaking it for, nor trying to turn it into something that is alien to its nature.

—Carl von Clausewitz, *On War*

Before you can fight Fourth Generation war successfully, you need to understand it. Because it is something new to our time, no one understands it completely. It is still evolving and taking shape, which means our understanding must continue to evolve as well. This chapter lays out our best current understanding of the Fourth Generation of modern war.

The Three Classic Levels of War

The three classic levels of war—strategic, operational, and tactical—still exist in Fourth Generation war. But all three are affected, and to some extent changed, by the Fourth Generation. One important change is that, while in the first three generations strategy was the province of generals, the Fourth Generation has given us the "strategic corporal." These days, the actions of a single enlisted man can have strategic consequences, especially if they happen to take place when cameras are rolling.

The Second Persian Gulf War provides numerous examples. In one case, U.S. Marines had occupied a Shi'ite town in southern Iraq. A Marine corporal was leading a patrol through the town when it encountered a funeral procession coming the other way. The corporal ordered his men to stand aside and take their helmets off as a sign of respect. Word of that action quickly spread around town, and it helped the Marines' effort to be welcomed as liberators. That in turn had a strategic impact, because the American strategy depended upon keeping Shi'ite southern Iraq quiet in order for American supply lines to pass through the territory.

Another change is that all three levels may show up in a small local theater of operations. A troop unit may have a beat, much as police do—an area where they are responsible for maintaining order and perhaps delivering other vital public services. The unit must harmonize its local tactical actions with higher strategic and operational goals, both of which must be pursued consistently on the local level. (Note: When a unit is assigned a beat, it is important that the beat's boundaries reflect real local boundaries, such as those between tribes and clans, and not be arbitrary lines drawn on a map at some higher headquarters.)

These changes point to another of the dilemmas that typify Fourth Generation war: what succeeds on the tactical level can easily be counter-productive at the operational and strategic levels. For example, by using their overwhelming firepower at the tactical level, state forces may in some cases intimidate the local population into fearing them and leaving them alone. But fear and hate are closely related, and if the local population ends up hating the state forces, that works toward their strategic defeat.

This is why in Northern Ireland, British troops were not allowed to return fire unless they were actually taking casualties. The Israeli military historian Martin van Creveld argues that one reason the British Army did not lose in Northern Ireland is that it accepted more casualties than it inflicted.

Fourth Generation war poses an especially difficult problem to operational art: put simply, it is difficult to operationalize. Often, Fourth Generation opponents have strategic centers of gravity that are intangible. These may involve proving their manhood to their comrades and local women, obeying the commandments of their religion, or demonstrating their tribe's bravery to other tribes. Because operational art is the art of focusing tactical actions on enemy strategic centers of gravity, operational art becomes difficult or even impossible.

This was the essence of the Soviet failure in Afghanistan. The Soviet Army, which focused on operational art, could not operationalize a conflict where the enemy's strategic center of gravity was God. The Soviets were reduced to fighting at the tactical level, where their army was not very capable, despite its vast technological superiority over the Afghan mujahideen.

Fourth Generation war affects all three classical levels of war. An example comes from Colonel John Boyd's definition of grand strategy, which is the highest level of war. He defined grand strategy as the art of connecting yourself to as many other independent power centers as possible, while at the same time isolating your enemies from as many other power centers as possible. A Fourth Generation conflict will usually have many independent power centers, not only at the grand strategic level but down all the way to the tactical level. The game of connection and isolation will therefore be central to tactics and operational art as well as to strategy and grand strategy. It is important to ensure that what you are doing at the tactical level does not alienate independent power centers with which you need to connect at the operational or strategic levels. Similarly, you need to be careful not to isolate yourself today from independent power centers to which you will need to connect tomorrow.

While the three classic levels of war carry over into the Fourth Generation, they are transformed. We do not yet know all the ways in which they will change when state-armed forces face Fourth Generation opponents, but we know they will change. As we gain experience in Fourth Generation conflicts, our understanding must grow as well. It is vital that we remain open to new lessons and not attempt to force the new forms of warfare we are observing into outdated conceptual frameworks.

Three New Levels of War

While the three classic levels of war carry over into the Fourth Generation, they are joined there by three new levels which may ultimately be more important. Colonel Boyd identified these three new levels as the physical, the mental, and the moral levels. Furthermore, he argued that the physical level—killing people and breaking things—is the *least* powerful, the moral level is the most powerful, and the mental level lies between the other two. Colonel Boyd argued that this is especially true in guerrilla warfare, which is more closely related to Fourth Generation war than is formal warfare between state militaries. The history of guerrilla warfare, from the Spanish guerrilla war against Napoleon through Israel's experience in southern Lebanon, supports Colonel Boyd's observation.

This leads to the central dilemma of Fourth Generation war: *what works for you on the physical (and sometimes mental) level often works against you at the moral level.* It is therefore very easy to win all the tactical engagements in a Fourth Generation conflict yet still lose the war. To the degree you win at the physical level by utilizing firepower that causes casualties and property damage to the local population, every physical victory may move you closer to moral defeat, and the moral level is decisive.

Some examples from the American experience in Iraq help illustrate the contradiction between the physical and moral levels:

1. The U. S. Army conducted many raids on civilian homes in areas it occupied. In these raids, the troops physically dominated the civilians. Mentally, they terrified them. But at the moral level, breaking into private homes in the middle of the night, terrifying women and children, and sometimes treating detainees in ways that publicly humiliated them (like stepping on their heads) worked powerfully against the Americans. An enraged population responded by providing the Iraqi resistance with more support at each level of war, physical, mental, and moral.

2. At Baghdad's Abu Ghraib prison, MPs and interrogators dominated prisoners physically and mentally—as too many photographs attest. But when that domination was publicly exposed, the United States suffered an enormous defeat at the

moral level. Some American commanders recognized this when they referred to the soldiers responsible for the abuse as, "the jerks who lost us the war."

3. In Iraq and elsewhere, American troops (other than Special Forces) quickly establish base camps that mirror American conditions: air conditioning, good medical care, plenty of food and pure water. The local people are not allowed into the bases except in service roles. Physically, the American superiority over the lives the locals lead is overwhelming. Mentally, it projects the power and success of American society. But morally, the constant message of "we are better than you" works against the Americans. Traditional cultures tend to put high values on pride and honor, and when foreigners seem to sneer at local ways, the locals may respond by defending their honor in a traditional manner—by fighting. After many, if not most, American military interventions, Fourth Generation war has tended to intensify and spread rather than contract.

The practice of a successful Fourth Generation entity, al-Qaeda, offers an interesting contrast. Osama bin Laden, who came from a wealthy family, lived for years in an Afghan cave. In part, this was for security. But bin Laden's choice also reflected a keen understanding of the power of the moral level of war. By sharing the hardships and dangers of his followers, Osama bin Laden drew a sharp contrast at the moral level with the leaders of local states, and also with senior officers in most state armies.

The contradiction between the physical and moral levels of war in Fourth Generation conflicts is similar to the tension between the tactical and strategic levels, but the two are not identical. The physical, mental, and moral levels all play at each of the three classic levels—tactical, operational, and strategic. Any disharmony among levels creates openings which Fourth Generation opponents will be quick to exploit.

War in Three Dimensions, Squared

Perhaps the best way to search out and identify potential disharmonies among these new and classic levels is to think of two intersecting games of three-dimensional chess. A single game of three-dimensional chess is challenging enough, in terms of the possible moves it offers. Now, imagine a single three-level game, representing the three classical levels of war, with another three-level game slashing through it at an angle. The second game represents Boyd's levels of war, the physical, the mental, and the moral. The complexity and the demands it makes on decision makers are daunting. But it is in just such a complex atmosphere that practitioners of Fourth Generation war must try to identify and avoid disharmonies among levels.

Fortunately, there is a simple tool that can help meet this complex challenge: the grid.

The Grid

	Physical	Mental	Moral
Tactical			
Operational			
Strategic			

By using the grid to evaluate every proposed mission before it is undertaken, it is often possible to avoid the sorts of contradictions and unwanted second-order effects that bedevil actions by state militaries in Fourth Generation wars. How can you do that? By asking what the effects of the mission are likely to be in each of the nine boxes.

Let's consider three examples, looking just at the basics.

First, killing the enemy physically reduces the threat, mentally it makes some potential enemies afraid to fight us, but morally it turns us into Goliath and may also obligate the relatives of those we kill to fight us. Going down, it counts as a win tactically, offers little but attrition operationally and works against us strategically because every fight is an escalation that diminishes the order that state forces are trying to restore. Since a higher level dominates a lower, on both scales killing the enemy is a net negative.

The second option is capturing the enemy. Physically, capturing the enemy is harder and more complicated than killing him. Mentally, it may be less frightening and thus less effective. But morally it works in our favor because the strong appear merciful (so long as prisoners are treated well), and a suspicion of cowardice hangs over anyone who surrenders. Looking down, a capture is equal tactically to a kill as a win, operationally it is still just attrition but strategically it is a plus because captives are useful chips in bargaining de-escalatory deals.

Net result: missions should put a premium on capturing enemy combatants rather than killing them.

Third, let's look at the operational level. How might our grid help us evaluate moving out of Forward Operating Bases and into villages, towns, and cities? Physically, the risk to our troops goes up. Mentally, we may be more apprehensive, but the people become less frightened of us as they get to know us. Morally, it is a huge plus because we are now protecting the people instead of living in isolation in order to protect ourselves. Going down: tactically, we may have to suffer more casualties than we inflict in order to de-escalate, which puts high demands on the self-discipline of the troops; operationally, it is a plus because when we establish order locally we are serving the commander's intent; and strategically, the spread of order is what leads to mission accomplishment and our return home.

As the boxes fill as we evaluate many potential missions, we begin to be able to do what John Boyd called many-sided cross-referencing. Of course, in considering the grid we must never forget the intent and the *Schwerpunkt* (focal point of effort), which are the first touchstones for any mission evaluation.

Fourth Generation War is Not Easy

Because war draws forth the ultimate in human powers, it is also the most complex of human activities. War is not a football game, nor is it merely an expanded version of a fistfight on the school playground. Because Fourth Generation war involves not only many different players, but many different kinds of players, fighting for many different kinds of goals (anything from money to political power to religious martyrdom), it is more complex than war between state militaries. Attempts to simplify that complexity by ignoring various elements merely set us up for failure. The worst possible simplification is reducing the problem to putting firepower on targets.

Scenario One: "Operation Goliath"

For General Braxton Butler's 13th Armored Division, the invasion of Inshallahland had been a cakewalk. Inshallahland's small air force had been destroyed on the ground in the first few hours. Apaches had knocked out most of the Inshallan tanks before his M-1s even saw them. Virtually all had been abandoned before they were hit. It seemed the Inshallan army just didn't have much fight in it. The 13th Armored Division swept into Inshallahland's capital in less than a week, suffering only a handful of casualties in the process. The local government skipped the country, taking the treasury with them, and an American proconsul now governed in their place. American-imposed secular democracy and capitalism would soon give the people a better life, or so General Butler thought.

But that is not quite how it turned out. Within days of the decisive American victory, graffiti began showing up, posting the message, "*Now the real war starts.*" It seemed those Inshallan soldiers who skedaddled so fast had taken their light weapons with them. Some analysts said that was the Inshallan strategy from the outset, although General Butler didn't pay much attention to eggheads like that. His job was to put steel on target.

So as the insurgency spread, that is what General Butler did. He called it "Operation Goliath." He knew no enemy on earth could stand up to American firepower. All that was necessary was killing anyone who resisted and scaring everyone else into cooperating with the Americans.

In town after town in the 13th Armored Division's sector, his troops methodically launched cordon-and-search operations. He kept his casualties down by prepping each town thoroughly, using air and artillery to take out any likely targets. Then his tanks and Bradleys swept through. He was certain he was killing a lot of bad guys; that much firepower had to do something. It made a mess of the towns, but fixing them was someone else's problem. Anyway, he was rotating home next week. In the meantime, Operation Goliath would clean out the town of Akaba.

Mohammed lived in Akaba. He was a poor man, like almost everyone in Akaba. But his tea shop across from the mosque allowed him to feed his family. He was even able to save some money so that someday he could go on the Haj.

When the 13th Armored Division troops first came through Akaba months before, Mohammed had watched. There wasn't any fighting, thanks be to Allah, but the American tanks had ripped up some roads, crushed sewers and water pipes, and even knocked down a few buildings. An American officer had promised they would pay for the damage, but they never did. Still, life went on pretty much as before. No one collected taxes now, which was good. Some foreigners, not Americans, Mohammed thought, had set up a medical clinic; they were welcome. The electricity was on more often, which was also good. Anyway, the Americans would leave soon, or so they said.

Of course, the mujahideen were now active in Akaba, as they were everywhere. Mostly, they set bombs by the sides of roads, targeting American supply convoys. He had seen an American vehicle burn after it was hit. He felt sorry for the American soldiers in the burning truck. They were someone's sons, he thought. War was bad for everyone.

When the bombing started in the night, Mohammed did not understand what was happening. Huge explosions followed, one after another. He quickly got his family out of the rooms over the tea shop where they lived and into the mosque across the street. He did not know who was doing the bombing, but perhaps they would not bomb a mosque.

At daybreak, the bombing stopped and American tanks came down his street. This time, they did not just pass through. American soldiers were kicking in the doors of every building and searching inside. The Americans were attacking the mujahideen. He knew some of the mujahideen. They were poor men, like him. They had few weapons. The Americans wore armor and helmets. Their tanks were enormous, and from the door of the mosque he could see their helicopters overhead, shooting anyone on the streets. Butchers! Murderers! How could human beings do this?

An American tank stopped near his tea shop. Suddenly, two mujahideen, just boys, ran out from the alley by his shop. They had a rocket-propelled grenade, or RPG. Before they could fire, the Americans' machine guns cut them down. By God, what an awful sight! Then the tank swiveled its enormous gun. It fired right through his shop into the alley. His business and his home were destroyed in an instant. "God curse them! God curse them!" Mohammed wailed. In less than a minute, he had lost his home and his livelihood.

American soldiers came into the mosque. They kept their boots on, defiling the holy place. They were screaming in a language Mohammed did not understand. His wife and children were terrified of the soldiers. In their helmets and armor and sunglasses, they looked like *djinn*, not men. Mohammed pulled his family into a corner and stood in front of them to protect them. He was a small man and had no weapon, but his honor demanded he defend his family. He could do nothing else.

Three American soldiers ran up to him, still screaming. He did not know what they wanted. Two of them grabbed him suddenly and threw him on the ground. One put his boot on Mohammed's head to hold him still. Enraged by the terrible insult and the humiliation in front of his family, Mohammed struggled. Another soldier kicked him in the groin as he lay on the ground. Retching with pain, he watched as the Americans ran their hands over his wife and daughter. They did something with his hands too. He did not know what. Then, without warning or explanation, they abruptly let him go and left him lying on the ground, frightened, hurt, humiliated, and angry.

Back in the 13th Armored Division's headquarters, General Butler's replacement had arrived. Major General Montgomery Forrest was invited by General Butler to join the brief on the progress of Operation Goliath. "Yesterday was another major success," General Butler told his replacement. "We pacified the town of Akaba, killing at least 300 muj and capturing 17. We've got a pretty good template for how to handle these places, and I don't think you'll have any problem picking up where I've left off."

That same day, Mohammed and his family were approached by Rashid. Mohammed knew Rashid was a mujahid. "We are sorry for what the American devils did to you yesterday," Rashid said. "My cousin said you and your family are welcome to live in his home. Here are 5,000 dinars to help you. We will also help rebuild your home and shop when the Americans have been driven out, God willing."

"Praise be to God for your generosity," Mohammed replied. "I want to fight the Americans too. But I am not a soldier. I saw how the American tank killed those two boys by my shop. The dogs even ran the tank over their bodies. You must have suffered many dead yesterday."

"Actually, praise be to God for his protection, we only had eleven men killed. The two you saw martyred were new to us. We told them to run away, to be safe until we could train them. But they took a weapon and attacked anyway. Now they are with God. But if you will join us, Mohammed, we will not throw your life away. We will train you well, so that when you fight the Americans you will kill many of them before you are made a martyr yourself. And we take care of our martyrs' families, so you will not need to worry about them. Thanks to the faithful, we have plenty of money, and weapons too."

"Do you know what the American dogs did?" Mohammed said. "They put their boots on my head, in front of my family. By God, I will fight them. I will be a suicide bomber myself."

Mohammed's son, who had just turned 13, had been listening to the conversation. "Father, I want to avenge our family's honor too. I want to be a suicide bomber also. Once I took candy from the Americans. Now I hate them more than I fear death."

"My son, if you had said this to me the day before yesterday, I would have beaten you. Now I give you my blessing. Go with Rashid and do whatever he tells you. Perhaps God will allow us to be martyrs together."

Scenario Two: "Operation David"

A week later, General Butler had moved on to his important new job at the U.S. Army's Training and Doctrine Command (TRADOC), where he would oversee the development of counterinsurgency doctrine. The division staff had worked hard on their first brief for the new commanding general. With 714 PowerPoint slides, they would show him how Operation Goliath would pacify its next target, the town of Hattin. The general was seated in the first row, coffee cup in hand. But before the briefer could begin, a lieutenant colonel in the seventh row of horse-holders stood up.

"General Forrest, before this brief starts, I have something I'd like to say."

Every head swiveled. Who was this guy interrupting the brief?

"Colonel, I apologize, but I'm so new here I'm afraid I have to ask who you are," General Forrest replied.

"I'm Lieutenant Colonel Ed Burke, sir, commander of 3rd Battalion, 13th Armored Division. Hattin is in my sector. Sir, I apologize for interrupting the briefing, but I've got something I have to get off my chest."

"Don't worry about the damn briefing," General Forrest replied. "Personally, I hate PowerPoint." The staff's sphincters tightened in unison. "What have you got to say?"

"Sir, I respectfully request that Operation Goliath not be carried out in Hattin."

"Why not?"

"Because it will make the situation there worse, sir, not better. I'm not saying we don't have problems in Hattin. We do. But while we don't have a 100 percent solution to the insurgency there, we have maybe a 51 percent solution. Operation Goliath represents the opposite of everything we've been doing. In my personal opinion, if operation Goliath hits Hattin, it will make our job there impossible. It will work for the resistance, not against it."

"This guy is toast," whispered one colonel on the staff to another.

"Well, I tend to think 51 percent solutions may be the best we can do against insurgents," said General Forrest. "Why don't you tell us what you're doing? Come on up front here and take over. The staff can just give me the briefing text and I'll read it over in my spare time."

"Thank you, sir," said LtCol Burke. "We call what we're doing in Hattin 'Operation David.' Sir, may I begin by asking the division staff how many casualties we have suffered in Akaba?"

The Division G-3 glared at Burke, but General Forrest looked like he expected an answer. "We have suffered five KIA since yesterday morning, with 23 wounded, 18 of whom had to be evacuated. Resistance is continuing for the moment, so I cannot say this will be the final casualty total. I expect all resistance will be crushed sometime tomorrow."

"Don't count on that," said General Forrest. "Please continue, Colonel."

"In Hattin, since my battalion took over four months ago, I have had two KIA and five wounded, all in two incidents. I have had only three successful attacks on American convoys in my whole sector, all by IEDs. As you know, General, metrics are pretty worthless in this kind of war. But as best we can tell, only one percent of the population in my sector is actively hostile. We believe we have caught everyone responsible for planting the IEDs that hit our convoys. We have captured over 1,000 insurgents. Most important, we have not killed a single Inshallan civilian."

"Excuse me, Colonel Burke," interrupted the G-3. "My records show you forwarded only 237 captured insurgents, not 1,000."

"That is correct, sir," replied LtCol Burke. "We release all the locals we capture. But first, we keep them with us for a while to show them what we are doing. They see with their own eyes that we are treating people with respect and trying to help. They also get to know my soldiers, whom I have ordered to treat detainees as guests of the battalion. Only if we capture someone a second time or if they are not from Hattin do we forward them to division as prisoners."

"Is this a 'hearts and minds' strategy, Colonel?" asked General Forrest.

"Not exactly, sir. We don't expect the locals to love us. We're foreign invaders and infidels to them. Our goal is to keep them from hating us so much that they fight us. I think we've done that pretty well, sir."

"Colonel, why don't you start from the beginning and tell us the whole story of Operation David?" asked General Forrest.

"Yes, sir. Well, when we knew where our sector was going to be, I gathered all my officers and senior NCOs, and some junior NCOs and troops as well, and told them what I wanted. I told them we had to operate in a way that would not make the locals hate us enough to fight us. Then I asked how we could do that. They talked, and I listened. I had an advantage in that we have a company of National Guardsmen attached. A lot of them are cops. I think cops understand this kind of situation better than a lot of soldiers do.

"The cops made one very important point right at the beginning.

They said the key to keeping the peace is to de-escalate situations rather than escalate them. Soldiers are taught to escalate. If something isn't working, bring in more firepower. Cops don't do that, because it enrages the community and turns it against them. So that was one piece of the puzzle. Another came from our battalion chaplain. He opened the Bible and read the story of David and Goliath. Then he asked how many of us were rooting for Goliath? My light bulb went on at that point, and I said what we want is Operation David.

"An NCO said that if we want to be David, we should just carry slingshots. Everybody laughed, but I saw his point. I said we won't go in with M-1s and Bradleys. Just HMMWVs and trucks. A private suggested we ditch the helmets, armor, and sunglasses because they make us look like Robocop. I decided he was right, so we did that too."

"Are you saying you aren't using all your assets?" the incredulous G-3 asked.

"That is correct, sir," LtCol Burke replied. "One of our first rules is proportionality. A disproportionate response, like using an M-1 tank against a couple lightly-armed mujahideen, is a great way to make the locals hate us so much they will fight us. It also makes us look like cowards."

"That sounds like you are taking unnecessary risks with American lives," the G-3 responded.

"Sir, how do we lose more American lives, by using our own infantry against their light infantry, or by turning on massive firepower that serves as our enemies' best recruiting tool? Sir, I have to wonder if you are missing the forest for the trees."

"Personally, I am more interested in the forest," said General Forrest. "Please continue, Colonel."

"Yes, sir. One of my National Guard officers served in Bosnia. He said the Europeans and the locals all laughed at us for hunkering down in fortified camps and seeming scared all the time. It's the old Force Protection crap. So I said, 'Can it. No Fort Apaches. We'll live in the towns. We will billet with the people, paying them well for the quarters we occupy. We'll shop in the local markets, drink coffee in the local cafes.' In Hattin, my headquarters is over a row of shops, right downtown. We protect the shopkeepers, but they also protect us. They don't want their shops blown up. I have troops living that way all over town. I let my captains, lieutenants, and sergeants work their areas the way they see fit, blending in as much as possible."

"With that kind of dispersion, how do you control your men?" asked the increasingly angry G-3.

"I don't," Burke shot back. "I believe in command, not control. I give my subordinates mission orders. They know the result I want, and I leave it up to them how to get it. If they need help, they come see me and we talk. Otherwise, I trust them to get the result. If one of them can't, I relieve him."

"Tell me about your KIA," General Forrest interjected.

"Yes, sir. It happened within the first couple weeks. A suicide bomber in a car hit one of my patrols. I lost two KIA and three wounded, all with limbs blown off. But 11 Inshallans were also killed and 32 wounded. I immediately ordered that we treat their wounded just like our own. We sent them on helos to American-run hospitals, not the crummy local ones. We transported their families to the hospitals to see them, and when they were well enough we brought them to their homes. We also gave money to the families that had lost wage-earners.

"Moslems bury their dead immediately, and I and my men went to all the funerals. Then I had memorial services for my two KIA and invited the townspeople. Many came, including three imams who offered prayers. That had a huge impact locally. I then asked the imams if they and their colleagues would give classes on Islam to me and my troops. That also had a huge impact, and it helped build my guys' cultural intelligence.

"Sir, my other two were wounded when a couple kids with AK-47s jumped one of my patrols. They couldn't really shoot, it was just spray-and-pray. Despite the two men down, my guys did not shoot the kids. My patrol leader charged them and they dropped their weapons and ran. When he caught them, he brought them back to the ambush site, pulled their pants down and spanked them. The crowd loved it, and the kids were humiliated in front of their buddies instead of being heroes. Both of my wounded guys have since returned to duty and the kids' parents have apologized to us. They were very grateful that we did not kill their sons."

"How did you train for this?" General Forrest asked.

"Well, sir, as one example, when I took my battalion through the 'local village' training stateside before we deployed, I reversed roles. I had my guys play the villagers, and I had troops who didn't speak their language sweep through on a typical cordon-and-search mission. I made sure the troops treated my villagers like we too often treat locals—screaming at them in a language they did not understand, throwing them around, detaining them in painful positions, and so forth. The result was just what I wanted—a lot of fights. My guys got so angry they started throwing punches. Then in the debrief I asked them, 'If we don't want the locals to fight us, how should we treat them?' The fact that they had been on the receiving end helped them see themselves in a whole new light."

"I think I might want to do that with my other units," General Forrest said. "Please continue."

"Yes, sir. From day one, our message to the people of Hattin was, 'We're not here to take over. You are in charge. You tell us what to do that will help you.' We helped them bring in NGOs to set up clinics and distribute food. We put our troops to work under the local Inshallan engineers and technicians to improve the infrastructure. I made my HQ a 'go to' point for the Inshallans when they needed parts or equipment. Over and over, we made the point that we are there to serve. On security, we let the mayor and the local police set policy. We only help when they ask us. They want order, which is what we want too, only they know a lot better than we do how to get it in their society."

"We understand that real psyops are not what we say but what we do, and God help us if the two are different. The people of Hattin now understand that we are not there to change the way they live, or to make them live by our rules. Hattin is a fundamentalist Islamic city, and some of their practices bother us. But this is their country, not ours. I've had signs put up in all our buildings, in Arabic and in English, that say, 'When in Inshallahland, do as the Inshallans do.' We go out of our way to make it clear that we do not see our way of life as superior to theirs. We are not somehow 'better' than they are. In cultures like this one, honor and pride are very important. If we seem to lord it over them, they have to fight us because their honor demands it."

"Stop for one minute, Colonel," interrupted the G-3. "We have similar humanitarian assistance programs as part of Operation Goliath. After we have secured a town, we bring in NGOs too. Do you know what the insurgents do to them? They capture them, hold them for ransom and then cut their heads off! Are you telling me that does not happen in Hattin?"

"Well, that brings us to the next level," replied LtCol Burke. "Life is harder for insurgents in Hattin than in the towns where Operation Goliath has left its heavy footprint. It is easy for insurgents in your towns to gain the people's support because Operation Goliath has made Americans hated, hated bad enough that lots of people want to see them killed. That is not true in Hattin. Why would people want to capture aid workers when they are just helping?"

"You did not answer the question," the G-3 pointed out. "Have any of your aid workers been captured?"

"Yes. Unfortunately, there will always be some people that we refer to as 'bad apples.' Operation David has kept their number small, but they exist. We have to deal with them in a very different way. We have to capture them or kill them."

"That's no different from what we do," said the G-3.

"Yes it is, because how we do it is different," Burke replied. "We never do cordon-and-search. We never kick down doors. We never terrorize civilians or call in heavy firepower. If we have to take some-one out, our preferred option is to take out a contract on them. Locals do the dirty work, and we leave no American fingerprints.

"If there is an insurgent cell that is too tough for locals to handle, we send in our Nighthunters, our equivalent to Delta Force. They are experts in low-impact combat. They specialize in being invisible. Local citizens never see them or deal with them. That enables us to keep the locals from seeing the average American soldier as a threat. Our cops put the Nighthunter concept together. It is like a SWAT team. People don't confuse SWAT with their local cop on the beat. Every time we've had an aid worker taken hostage, the Nighthunters have rescued them within 24 hours."

"Colonel Burke, I'm the Public Affairs Officer (PAO) on the 13th Armored Division staff," said a reservist. "How are you working the press problem in Operation David?"

"By playing one media operation off against others," Burke replied. "I thought from the beginning that we would get favorable media coverage of what we are doing in Hattin, and on the whole I've been right. 90 percent of what we do is open to any reporter who wants to come along. That includes Al Jazeera.

"Just once, early on, Al Jazeera did an unfair and inaccurate story on one of our operations. In response, instead of kicking them out of Hattin, I invited Al Arabiya in. I knew they were competitors. I encouraged Al Arabiya to do an investigative report on the operation Al Jazeera had portrayed negatively, and I opened all our records up to them. Their report showed that Al Jazeera had been wrong. Since then, Al Jazeera has been very careful to get their facts right in Hattin. And that's all I ask. If we do something wrong and they report it, that's our fault, not theirs."

"It sounds to me as if Operation David requires superb local intelligence," General Forrest said. "How do you obtain that intelligence?"

"The same way cops do, by talking to the local people all the time," Burke answered. "They talk to us. Remember, we haven't made ourselves hated. We buy from locals all the time. Good customers become friends, and friends pass information to other friends. The real problem is the language barrier. We've worked on that a number of ways. Of course, we've hired as many locals as interpreters as we can. I have them give classes each day to all my troops, so they learn at least some phrases and common courtesies in the local language. Each of my men has a pack of flash cards with basic phrases in English and Arabic, the Arabic spelled phonetically and also in script. If he can't say it right, at least he can point.

"Our Guardsmen have been a tremendous help in this regard. They come from Cleveland, Ohio, which has a large Arabic-speaking population. With the support of and funding from the State of Ohio, when they knew they were deploying here, they offered special one-tour enlistment packages, with big bonuses, to anyone in Cleveland who could speak Arabic. It didn't matter how old they were, there was no PFT, all they wanted was translators they knew would be loyal to us. Those guys are terrific.

"Finally, I've told the locals that anyone who works for us will be eligible for a Green Card when American forces leave Inshallahland. Frankly, General, I've gone out on a limb here. That promise has done more than anything else to give us the language capability we need, but I don't know how I am going to keep it."

"Let me work on that one," replied General Forrest. "I think that is a great idea, and I have some friends back in Washington who may be able to help us do that."

The Division G-2 had been listening intently to the discussion. "Have any of our intelligence systems been useful to you, Colonel?" he asked Burke.

"Yes and no," Burke replied. "I have to say that virtually all the intel we've received from higher has been either too late or wrong or both."

"That's no surprise to me," replied the G-2. "Our systems were all designed to collect and analyze data on other state militaries. What are our satellites supposed to do in this kind of war, watch a twelve-year-old boy pick up a stone?"

"But we have used technology effectively on the local level," Burke said. "We use our superb night vision capability to cover virtually all of Hattin at night. I have night observation posts (OPs) everywhere. With rare exceptions, all they do is observe and note patterns. We don't hassle people for being on the street at night. As any cop will tell you, safe streets have people on them, day and night. It is empty streets that are dangerous. If my guys see something going down, it's usually street crime, so they call the local cops. Of course, the locals know we are doing this—the locals know everything we do, often before I know it—but because we don't hassle them, it's OK. They want safety and order as much as we do.

"We have also emplaced small, camouflaged cameras and listening devices in some key places. I'd rather not go into too much detail as to how many and where. But I can say that there aren't many phone conversations in Hattin, or meetings in large spaces, that we are not aware of. All this information is available to any of my leaders who want it, right down to the squad level. It is an open-architecture intel system. We do not hoard intelligence in my HQ. I'm not a dragon who wants to sleep on a pile of gold."

The G-2 smiled. "If I could trade my eagles for captain's bars, I think I'd enjoy being your S-2," he said.

"Why don't you do that?" asked General Forrest. "See how they are making it work, then come back here and try to do the same thing for me."

"Roger that, sir," said the G-2. "Gee, I'll really miss all my computers. I might even get to see the sun!"

"You are welcome to come back with me and stay as long as you want," Burke said to the G-2. "Just be aware that our intel system, like everything else, is a flat network, not a hierarchy. My units pass intel laterally and down, not just up a chain. It's like German-style armor tactics, in that we are more reconnaissance-driven than intel-driven."

"That's how the tactical level has to work," said the G-2.

"Can you give me an example?" asked General Forrest.

"Easily, sir," Burke replied. "Let me come back to the G-3's question about kidnapping. The first time that happened, we immediately tapped our whole human intel network. The main way we did that was by having our guys go to the cafes and tea rooms and put out the word, which included a lot of cash for intel that proved good. Then I gathered all our squad and platoon leaders and asked them to game the situation. In a matter of hours we were sure we had the location, and when the Nighthunters went in, we were spot on. Of course, the fact that we were able to do that and do it fast sent a message to the insurgents and to the whole town, so the rescue had strategic as well as tactical consequences. It played on the physical and mental levels of war, and I think perhaps on the moral level as well, because even though we had to use violence, no innocents were harmed. In fact, as is usually the case in our Nighthunter ops, no one was killed."

"You didn't kill the enemy?" the G-3 interjected.

"No, sir, we try not to. Sometimes we can't avoid it, but in a clan- and tribe-based society like this one, if you kill somebody you have a blood feud with his relatives. Because the insurgents don't have gas masks, the Nighthunters usually flood the place with CS, then just walk in and round people up. We treat all the captives with respect, and when we do kill someone, we pay blood money to his family, clan, and tribe. We are always trying to de-escalate, sir, not escalate. We don't want to create martyrs for the other side.

"Of course, there are situations where we do want bloodshed. We constantly try to identify factional divisions among the insurgents. When we find one, we try to escalate it, to ramp up friction within the other side. We use lies and deceptions to bring one faction to the point where it wants to whack another, then we find discreet ways to help them do that. We do it in such a way that they all start blaming each other. Often, the insurgents do our most difficult jobs for us, killing their own leaders out of fear of being stabbed in the back. Remember, this isn't a culture that has much trust in it.

"One time, we planted someone to get kidnapped. He was a Nighthunter disguised as an NGO worker. We had implanted a tracking device in his body. During his captivity he was able to learn a lot about our enemies. It was easy to rescue him because we knew exactly where he was.

"We often spot people who are trying to bring weapons into Hattin or hide them there. We do not interrupt those operations. We don't try to capture or destroy those weapons. Instead, one of our Guardsmen knew of some stuff we could spray on their ammunition that they would not readily notice but would cause it to jam in the weapon. I had cases of the stuff in spray cans shipped in from Cleveland. We sneak in and spray their ammo stocks, then when they try something, their weapons don't work. That really undercuts their morale. If we seized or blew up their weapons, they could fight us by bringing in more or learning to hide them better. But they can't fight us because they don't know what we are doing. Their operations fall apart and they don't know why.

"They cannot ambush us because we follow no predictable patterns. They cannot surprise us because we are always watching, and they don't know when or where they are being watched. They cannot fight back without alienating their own people. All they see are the smiling faces of my men, who have now become part of their neighborhoods and communities."

"Anyway sir, that's Operation David. It's working in Hattin and in the rest of my sector. All I'm asking, sir, is please don't destroy everything we've worked so hard to build by having Goliath stomp on Hattin. There are plenty of other towns out there to wreck. Let Goliath go somewhere else."

"Well, Colonel, I think that is a reasonable request," said General Forrest. "I can tell you where Operation Goliath is going next. It is going in the wastebasket. Colonel Burke, I suspect Operation David could continue in Hattin without you for a while."

"Yes, sir, it could," Burke replied. "I didn't create Operation David and I don't run it. My men created it and they run it."

"Good, because I want you to come here, take over the G-3 shop for a while and expand Operation David to the whole 13th Armored Division. Can you do that?"

The lieutenant colonel thought for a few moments. "I think so, sir, if you will allow the men in the other battalions to do what mine have done."

"I will," said the general. "Meanwhile, I would like to ask my G-3 to go back to Hattin with your battalion, as an observer."

"Aye, aye, sir," responded the G-3, with a distinct lack of enthusiasm. He sensed that his moment might have come, and gone.

"One final request, Colonel Burke," said General Forrest. "Do you think you might present the division's Operation David to me without PowerPoint?"

"Yes, sir!" said Burke with a smile. "With your permission, I'd like to do with the division's PowerPoint stuff what I did with my battalion's."

"What is that, Colonel?" General Forrest asked.

"I let the insurgents capture it. It's slowed their OODA loop down to a crawl."

"Another good idea, Colonel," Forrest replied. "I always knew PowerPoint would be useful for something."

"Hot Wash"

If we critique Operation David, what lessons about Fourth Generation war do we see? First, we see elements that carry over from Third Generation (maneuver) warfare. They include:

- Outward focus. To have any hope of winning, a state military must focus outward on the situation, the result, and the action the situation requires, not inward on set rules, processes, and methods. Stereotyped tactics and all predictable patterns must be avoided. Commanders and units must be judged by the results they achieve, not whether they "go by the book."

- Decentralization. Authority and information flows, including intelligence dissemination, must be decentralized, often down through the most junior level of command and even to the individual level. This in turn requires trust both up and down the chain of command.

- Accuracy. Going through the OODA loop (Observe-Orient-Decide-Act) faster than your enemies remains important, but accuracy of Observation and Orientation may be more important than speed.[5] Because Fourth Generation forces are usually "flat," networked organizations, state-armed forces must "flatten" their own hierarchical structures in order to remain competitive.

In addition, this hypothetical example shows that in Fourth Generation war the moral level is dominant over not only the physical but also the mental level. Mentally, Mohammed thought he could not stand up to American technology, but the moral level compelled him to fight anyway.

We also see the power of weakness. In Fourth Generation warfare, the weak often have more moral power than the strong. One of the first people to employ the power of weakness was Mahatma Gandhi. Gandhi's insistence on non-violent tactics to defeat the British in India was and continues to be a classic strategy of Fourth Generation war. When the British responded to Indian independence rallies with violence, they immediately lost the moral war.

Operations David and Goliath show a strong military force, with almost no limits on the amount of violence it can apply to a situation, versus a very weak irregular force. The weaker force has the moral high ground because it is so weak. No one likes bullies using their physical superiority in order to win at anything, and unless we are extremely careful in how we apply our physical combat power, we soon come across as a bully, i.e. Goliath.

Most important, we see the central role of de-escalation. In most Fourth Generation situations, our best hope of winning lies not in escalation but in de-escalation (the "Hama model" discussed in the next chapter relies on escalation, but political factors will usually rule this approach out). De-escalation is how police are trained to handle confrontations. From a policeman's perspective, escalation is almost always undesirable. If a police officer escalates a situation, he may even find himself charged with a crime. This reflects society's desire for less, not more, violence. Most people in foreign societies share this desire. They will not welcome foreigners who increase the level of violence around them.

For state militaries in Fourth Generation situations, the policeman is a more appropriate model than the soldier. Soldiers are taught that, if they are not achieving the result they want, they should escalate: call in more troops, more firepower, tanks, artillery, and air support. In this respect, men in state-armed forces may find their own training for war against other state-armed forces works against them. They must realize that in Fourth Generation war, escalation almost always works to the advantage of their opponents. We cannot stress this point too strongly. State militaries must develop a "de-escalation mindset," along with supporting tactics and techniques.

There may be situations where escalation on the tactical level is necessary to obtain de-escalation on the operational and strategic levels. In such situations, state-armed forces may want to have a special unit, analogous to a police SWAT team, that appears quickly, uses the necessary violence, then quickly disappears. This helps the state servicemen with whom local people normally interact to maintain their image as helpful friends.

Proportionality is another requirement if state militaries want to avoid being seen as bullies. Using tanks, airpower, and artillery against lightly armed guerrillas not only injures and kills innocent civilians and destroys civilian property, it also works powerfully at the moral level of war to increase sympathy for the state's opponents. That, in turn, helps our Fourth Generation enemies gain local and international support, funding and recruits.

De-escalation and proportionality in turn require state-armed forces to be able to empathize with the local people. If they regard the local population with contempt, this contempt will carry over into their actions. Empathy cannot simply be commanded; developing it must be part of training. Both empathy and force protection are best served by integrating servicemen with the local population. If they live in a fortified base, separate from the local people, it will inevitably create a hostile us/them attitude on the part both of locals and the servicemen themselves. This isolates the state's armed forces from the local people, which works to the advantage of our opponents.

Empathy and integration permit effective cultural intelligence, which is to say, becoming able to understand how the society works. In Fourth Generation war, virtually all useful intelligence is human intelligence (HUMINT). Often, such HUMINT must both be gathered and acted on with stealth techniques, where the state's actions remain invisible to the local population. As in Third Generation war (maneuver warfare), the tactical level in Fourth Generation conflicts is reconnaissance-driven, not intelligence-driven. The information state militaries need will almost always come from below, not from higher-level headquarters.

An understanding of local, factional politics, including those within the camps of non-state opponents, will be of central importance to the effectiveness of operations. Success is more likely through leveraging such factionalism than through a force-based direct approach. We must understand that despite our vast tactical and technical superiority over most Fourth Generation opponents, at the strategic level, we will almost always be the weaker party. The reason is simple: at some point we will go home, while our opponents will remain. In the battle for the support of the local population, that fact overwhelms all others. Every local citizen must ask himself, "What will my situation be when foreign forces leave?" If we fracture the local society to the point where order is unlikely after we depart, anyone who has worked with us will be in danger.

Operation David illustrates a final central point about Fourth Generation war: our strategic objectives must be realistic. Seldom if ever will we be able to remake other societies and cultures. If doing so is our strategic objective, we will probably be doomed to defeat before the first round is fired. Nor can we make ourselves loved by countries we invade; keeping them from hating us so much that they want to fight us will often be the best we can do. In Fourth Generation wars, 51 percent solutions are acceptable.

Each of these points touches a central characteristic of Fourth Generation war. If we fail to understand even one of them, and act so as to contradict it, we will set ourselves up for defeat. Remember, for any state military, Fourth Generation wars are easy to lose and very challenging to win. This is true despite the state military's great superiority over its Fourth Generation opponents at the physical level of war. Indeed, to a significant degree, it is true because of that superiority. In most Fourth Generation wars, state-armed forces end up defeating themselves.

Chapter 2

Fighting Fourth Generation War

> *Without changing our patterns of thought, we will not be able to solve the problems we created with our current patterns of thought.*
>
> —Albert Einstein

At this point, you should have some understanding of Fourth Generation war. In this chapter, we will discuss how the armed forces of the state should fight in Fourth Generation conflicts.

Preparing for War

In Book Two, Chapter Two of *On War*, Clausewitz draws an important distinction between preparing for war and the conduct of war. Most of this chapter will be devoted to the conduct of Fourth Generation war. But there are some preconditions that fall under "preparing for war" we must address first. If these preconditions are not met, success is unlikely.

The first precondition is officer education and training that produces adaptive leaders. The schools must constantly place students in difficult, unexpected situations, then require them to make decisions and take action under time pressure. Schools must take students out of their comfort zones. Stress—mental and moral as well as physical—must be constant. War games, map exercises, and free-play field exercises must constitute the bulk of the curriculum. Drill and ceremonies are not important. Higher command levels overseeing officers' schools must learn to view high drop-out and expulsion rates as indications that the job of preparing new officers is being done correctly. Those officers who successfully graduate from the schools must continue to be developed by their commanders. Learning cannot stop at the schoolhouse door.

The second precondition is developing a viable strategy before entering a Fourth Generation conflict. We have already noted that our strategic goals must be realistic; we cannot remake other societies and cultures in our own image. Here, we offer another warning, one related directly to fighting Fourth Generation war: our strategy must not be so misconceived that it provides a primary reason for others to fight us.

Unlike state-armed forces, most Fourth Generation entities cannot simply order their men to fight. Most Fourth Generation fighting forces are, in effect, militias. Like other militias throughout history, motivating them to fight is a major challenge. We must always be careful to ensure we do not solve that problem for Fourth Generation opponents by adopting a strategy that makes their militiamen want to fight us.

What that means in specific situations varies case-to-case. And, the rule of not providing the enemy's motivation applies to operational art and tactics as well as strategy. We emphasize the strategic

level here in part because errors at the strategic level cannot be un-
done by successes at the operational and tactical levels. That is the
primary lesson from Germany's defeats in both World Wars; a higher
level of war trumps a lower. States often violate this rule in Fourth
Generation conflicts, and when they do so, they are defeated.

Fighting Fourth Generation War: Two Models

In fighting Fourth Generation war, there are two basic approaches or
models. The first may broadly be called the "de-escalation model,"
and it is the focus of this handbook. But there are times when state-
armed forces may employ the other model. Reflecting a case where
this second model was applied successfully, we refer to it (borrowing
from Martin van Creveld) as the "Hama model." The Hama model
refers to what Syrian President Hafez al-Assad did to the city of Hama
in Syria when a non-state entity there, the Moslem Brotherhood, re-
belled against his rule.

In 1982, in Hama, Syria, the Sunni Moslem Brotherhood was
gaining strength and was planning on intervening in Syrian politics
through violence. The dictator of Syria, Hafez al-Assad, was alerted by
his intelligence sources that the Moslem Brotherhood was looking to
assassinate various members of the ruling Ba'ath Party. In fact, there
is credible evidence that the Moslem Brotherhood was planning on
overthrowing the Shi'ite/Alawite-dominated Ba'ath government.

On February 2, 1982, the Syrian Army was deployed into the area
surrounding Hama. Within three weeks, the Syrian Army had completely devastated the city, resulting in the deaths of between 10,000
and 25,000 people, depending on the source. The use of heavy artillery, armored forces, and possibly even poison gas resulted in large-
scale destruction and an end to the Moslem Brotherhood's desires to

overthrow the Ba'ath Party and Hafez al-Assad. After the operation was finished, one surviving citizen of Hama stated, "We don't do politics here anymore, we just do religion."

The results of the destruction of Hama were clear to the survivors. As the June 20, 2000, *Christian Science Monitor* wrote, "Syria has been vilified in the West for the atrocities at Hama. But many Syrians—including a Sunni merchant class that has thrived under Alawite rule—also note that the result has been years of stability."

What distinguishes the Hama model is overwhelming firepower and force, deliberately used to create massive casualties and destruction, in an action that ends quickly. Speed is of the essence to the Hama model. If a Hama-type operation is allowed to drag out, it will turn into a disaster on the moral level. The objective is to get it over with so fast that the effect desired locally is achieved before anyone else has time to react or, ideally, even to notice what is going on.

This handbook devotes little attention to the Hama model because situations where the Western states' armed forces will be allowed to employ it will probably be few and far between. Domestic and international political considerations will normally tend to rule it out. However, it could become an option if a Weapon of Mass Destruction were used against a Western country on its own soil.

The main reason we need to identify the Hama model is to note a serious danger facing state-armed forces in Fourth Generation situations. It is easy, but fatal, to choose a course that lies somewhere between the Hama model and the de-escalation model. Such a course inevitably results in defeat, because of the power of weakness.

The military historian Martin van Creveld compares a state military that, with its vast superiority in lethality, continually turns its firepower on poorly-equipped Fourth Generation opponents to an adult who administers a prolonged, violent beating to a child in a

public place. Regardless of how bad the child has been or how jus-
tified the beating may be, every observer sympathizes with the child.
Soon, outsiders intervene, and the adult is arrested. The power mismatch is so great that the adult's action is judged a crime.[6]

This is what happens to state-armed forces that attempt to split the
difference between the Hama and de-escalation models. The seem-
ingly endless spectacle of weak opponents and, inevitably, local civilians being killed by the state military's overwhelming power defeats
the state at the moral level. That is why the rule for the Hama model
is that the violence must be over fast. It must be ended quickly! Any
attempt at a compromise between the two models results in prolonged
violence by the state's armed forces, and it is the duration of the mismatch that is fatal. To the degree the state-armed forces are also foreign invaders, the state's defeat occurs all the sooner. It occurs both
locally and on a global scale. In the 3,000 years that the story of David
and Goliath has been told, how many listeners have identified with
Goliath?

In most cases, the primary option for state-armed forces will be
the de-escalation model. What this means is that when situations
threaten to turn violent or actually do so, state forces in Fourth Generation situations will focus their efforts on lowering the level of con-
frontation until it is no longer violent. They will do so on the tactical,
operational, and strategic levels. The remainder of this handbook is
therefore focused on the de-escalation model for combatting insurgency and other forms of Fourth Generation warfare.

Less is More

When a state's armed service is given a mission to intervene in a Fourth
Generation conflict, its first objective must be to keep its own foot-

print as small as possible. This is an important way to minimize the contradiction between the physical and moral levels of war. The smaller the state's physical presence, the fewer negative effects it will have at the moral level. This is true not only for intervening state forces but for the state they are attempting to buttress against Fourth Generation opponents as well.

If the situation is such that foreign troops' presence must be obvious—that is, we cannot limit it in *extent*—another way to minimize their footprint is to limit its *duration*. Therefore, state-armed forces will often attempt to deal with Fourth Generation enemies by conducting punitive expeditions or raids instead of occupying an area. These raids will usually be sea-based. If all else fails, only then should a state resort to invading and occupying another country, usually as part of a joint or combined force. This is the least desirable option, because as foreign invaders and occupiers, the intervening forces will be at a severe disadvantage from the outset at the moral level of war.

Preserving the Enemy State

In situations where the armed forces of Great Powers invade and occupy another country, they will often find it relatively easy to defeat the opposing state and its armed forces. While this is a decisive advantage in wars between states, in Fourth Generation situations it brings with it a serious danger. In a world where the state is growing weaker, a victory can easily destroy the enemy state itself, not merely bring about regime change. If this happens, it may prove difficult or impossible to re-create the state. The result will then be the emergence of another stateless region, which is greatly to the advantage of Fourth Generation entities. As is so easily accomplished in Fourth Generation conflicts, the stronger side will have lost by winning.

Therefore, Great Powers must learn how to preserve enemy states at the same time that they defeat them. The specifics will vary according to the situation. But in many cases, the key to preserving the enemy state will be to preserve its armed forces.

Here, the revival of an 18th century practice may be helpful: rendering the opposing armed forces the "honors of war." Instead of humiliating them, destroying them physically or, after a Great Power victory, disbanding them, the winner should do them no more damage than the situation requires. Prisoners should be treated with respect. If they are senior officers, they should be treated as honored guests, invited to dine with the victor's generals, given the best available quarters and so forth. After a truce or armistice, the victor should praise how well they fought, give them every public mark of respect, and perhaps, through the next government, increase their pay. Throughout the conflict, the invader's actions should be guided by the goal of enabling and encouraging the local armed forces to work with the victors when it is over to preserve the state.

The same is true for the civil servants of the enemy state. It is critical that the state bureaucracy continue to function. Again, an immediate pay raise may be helpful. When foreign forces have to remove senior leaders of the state, the number removed should be as small as possible. The victor must be careful not to leave any segments of the enemy's society unrepresented in a new government. That government should be headed by local figures, not by someone from another country.

These matters will usually be decided at a level higher than the armed forces. But it is essential that senior officers speak forcefully to the political level about the need to preserve the enemy state after it is defeated. If that state disappears, the inevitable strengthening of Fourth Generation forces that will result will fall directly on the occupiers at the tactical level. Strong words from senior officers early can save many lives later. Taking the risks involved in offering such advice is part of the moral burden of command.

Fourth Generation Warfare is Light Infantry Warfare

As Fourth Generation war spreads, it is inevitable that, even if all the cautions offered above are heeded, Great Powers' armed forces will find themselves fighting Fourth Generation enemies. It is important both for the preparation for war and the conduct of war that they know that Fourth Generation war is, above all, light infantry warfare.

As a practical matter, the forces of most of our non-state, Fourth Generation adversaries will be predominantly irregular light infantry. Few Fourth Generation non-state actors can afford anything else, and irregulars enjoy some important advantages over conventional forces. They can be difficult to target, especially with air power and artillery. They can avoid stronger but more heavily equipped opponents by using concealment and dispersal, often within the civil population. They can fight an endless war of mines and ambushes. Because irregulars operate within the population and are usually drawn from it, they can solicit popular support or, if that does not work, compel popular submission.

Light infantry is the best counter against irregulars because it offers three critical capabilities. First, good light infantry, unless badly outnumbered, can usually defeat almost any force of irregulars it is likely to meet. It can do this in a man-to-man fight that avoids the "Goliath" image. If the light infantry does not burden itself too heavily with arms and equipment, it can enjoy the same foot mobility as the irregulars, enhanced as necessary by helicopters or attached motor vehicles.

Second, when it uses force, light infantry can be far more discriminating than other combat arms and better avoid collateral damage. Why? Because light infantry tactics do not depend on massive firepower. This is critically important at both the mental and moral levels.

Third, unlike soldiers who encase themselves in tanks or other armored boxes, fly overhead in tactical aircraft or man far-away artillery pieces or monitoring stations, light infantrymen can show the local population a human face. They can be courteous and even apologize for their mistakes. They can protect the local people from retaliation by the irregulars, assist with public works projects, or help form and train a local defense force.

U.S. military officers reading this handbook may think at this point that they are ahead of the game because they have light infantry in their force structure already. Unfortunately, what the U.S. military calls light infantry is really mechanized or motorized infantry without armored fighting vehicles (traditionally called "line infantry"). It possesses neither the tactical repertoire nor the foot mobility of true light infantry. A detailed discussion of the changes required to create genuine light infantry is found in Chapters 7–9. Here, we will note only that without true light infantry, states will seldom be able to come to grips with the elusive irregulars who will be their opponents in most Fourth Generation conflicts.

Out-G'ing the G: Lessons from Vietnam

Fourth Generation war is guerrilla warfare more than it is "terrorism." "Terrorism" is a special operation, a single tactical action designed to have direct operational or strategic effect. Because targets that have direct operational or strategic value are few and are usually well-protected, "terrorism" normally plays a minor role in Fourth Generation conflicts, although when it is successful the effects can be wide-ranging.

Most of what state militaries will face in Fourth Generation situations is a form of guerrilla warfare. Here, lessons from past guerrilla wars, especially Vietnam, remain relevant on the tactical level. Perhaps the most important lesson is that to defeat guerrillas, state-armed forces have to become better at their own game than they are. When U.S. Army Colonel David Hackworth commanded a battalion in the Vietnam War, he called this "out-guerrilla'ing the guerrilla," or "out-G'ing the G." In his memoirs, *About Face*, he wrote,

> *We would no longer be the counterinsurgents who, like actors on a well-lit stage, gave all their secrets away to an unseen, silent and ever-watchful (insurgent) audience in a darkened theater. Instead we would approach the battlefield and the war as our enemy approached it, and in so doing begin to outguerrilla the guerrilla—"out-G the G," as I hammered it again and again into the men of the Hardcore (battalion)—and win.*
>
> *The basic concepts behind my changes were that men, not helicopters or mechanical gimmicks, won battles, and that the only way to defeat the present enemy in the present war at a low cost in friendly casualties was through adopting the enemy's own tactics, i.e., "out G-ing the G" through surprise, deception, cunning, mobility … imagination, and familiarity with the terrain....* [7]

In training a military unit for Fourth Generation war, commanders should make use of the extensive literature on guerrilla warfare, from the Spanish guerrilla war against Napoleon to the present. Field training should be free-play exercises against guerrilla opponents who are allowed to make full use of such typical guerrilla tools as mines, booby traps and infiltration of their enemy's rear areas.

Integrating with the Local Population

Force Protection of the kind usually employed by American forces is highly disadvantageous in Fourth Generation war, because it seeks security by isolating the troops from the surrounding population. Effectiveness against Fourth Generation opponents demands the opposite: close integration with the local populace. Instead of making state forces less secure, integration will improve their security over the long run. The reason is that just as troops protect the local people, so the local people will protect them.

Perhaps the best example of this symbiotic protection is the traditional British policeman known as the "bobby." The bobby was, until recently, unarmed. The reason he did not need a weapon was that just as he protected the neighborhood, the neighborhood protected him. The bobby had a regular beat, which he patrolled on foot. He came to know every house and its inhabitants, and they came to know him. He became part of the neighborhood. Just as his familiarity with his beat enabled him to see very quickly if anything was out of the ordinary, so the fact that the local people knew him as an individual meant they told him what he needed to know. They did not want any harm to come to "their" bobby.

State forces will not be able to go about unarmed in most Fourth Generation situations. But they can become part of a neighborhood. To do so, they must live in that neighborhood, get to know the people who inhabit it and become known by them in turn. They will usually do so in small groups, squads, or even fire teams. To be effective, they must reside in the same neighborhood or village for some time. Results in Fourth Generation war usually come slowly.

American Marines had a program of integration with the local population during the Vietnam War, the Combined Action Program (CAP). By all accounts, it was highly effective. Commanders should attempt to learn from such past successes as the CAP program and not have to reinvent the wheel in each new conflict.

De-escalation

Unless state-armed forces are employing the "Hama model," it will be of decisive importance for them to manage most confrontations by de-escalating the conflict rather than escalating it. What does this require? First, they must understand that much of their training for combat is inappropriate. In most training, servicemen are taught that if they are not getting the result the situation requires, they should escalate. What this means is that their instincts will often be counter-productive in Fourth Generation conflicts. They must be conscious of this fact, or those instincts will drive them to escalate the level of violence and subsequently lose at the more important moral level of war.

Second, state militaries need to learn from police. The most common and most effective tool police use to de-escalate situations is talk. Here, foreign forces in Fourth Generation wars immediately find themselves at a disadvantage: they do not speak the local language. Nonetheless, they must develop ways to talk with the local population, including their opponents. Specific techniques are beyond the scope of a handbook. However, examples include:

- Hiring locals as interpreters. Always remember that locals who work with foreign forces must survive after the foreigners leave, which means they may have to work for both sides. A program where they could be offered a "Green Card" in return for loyal service could prove useful.

- Bringing citizens of the occupying country who are fluent in the local language into the military on a lateral-entry, no-military-training basis to provide interpreters whose loyalty can be trusted.

- Giving occupying forces "flash cards" with key words. The cards should include phonetic pronunciations; not all locals will be literate. Cards containing local gestures could also be useful.

In general, the key to successful communication is patience. Even with no common language, people can often communicate in a variety of ways. What is not useful is resorting to words screamed in a foreign language.

Perhaps the most important key to de-escalation is simply not wanting to fight. In April 2004, when U.S. Marines ended their first attempt to storm Fallujah in Iraq, the 1st Marine Division's commander, General Mattis, said, "We did not come here to fight." In Fourth Generation situations, that should be true in most encounters occupiers have with local people, including many armed Fourth Generation entities. Given the mismatch between occupying troops and local armed elements, any fighting works to the occupiers' disadvantage on the all-important moral level. In addition, the societal disorder fighting inevitably creates works to the advantage of non-state elements.

State-armed forces need to educate and train themselves to develop a mental switch. When the switch is set for combat with the armed forces of other states, they must want to fight. When instead it is set for Fourth Generation situations, they must be equally keen *not* to fight. Both objectives involve risks. But the second objective is just as important as the first, because *not* wanting to fight is as important for victory in the Fourth Generation as wanting to fight was in the Third.

One aspect of "not wanting to fight" may prove especially difficult for state militaries: in the Fourth Generation war, victory may require taking more casualties than you inflict. In most Fourth Generation situations, it is more important *not to kill* the wrong people than it is to kill armed opponents. This means that even when state forces are under fire, they must be trained and disciplined to return fire only when they are *certain* they are firing on armed enemies and on them *only*. Any time an innocent person is wounded or killed by foreign troops, his family and clan members may be required by the local culture to take revenge. Whenever that happens, Fourth Generation opponents are likely to get a stream of new recruits.

If state forces are fired on in a situation where it is not clear who is firing or those attacking are intermixed with the civilian population, the best solution may be to withdraw. State forces can later attempt to engage the enemy on their own terms. They need not "win" every firefight by leaving behind a pile of dead local people. In Fourth Generation conflicts, such "victories" are likely to add up to strategic defeat.

Finally, despite a policy of de-escalation, there will inevitably be situations where state forces do need to escalate. When that happens, we again stress that it must be over fast. To return to Martin van Creveld's analogy, an adult can get away with giving a kid one good whack in public. He cannot administer a prolonged beating. Once the escalation terminates, state forces must make every effort to demonstrate that de-escalation remains their policy.

Politics is War, and All Politics is Local

Clausewitz, writing of war between states, said that "War is the extension of politics by other means." In Fourth Generation situations, the opposite is more likely to be true: politics is an extension of war. This is consistent with the de-escalation model. Nowhere more than in a post-state, Fourth Generation situation is the old saying true, "All politics is local." When the state vanishes, everything becomes local. By understanding and leveraging local political balances, state forces may be able to attain many important objectives without fighting.

A useful model here is the old British Northwest Frontier Agent. The Northwest Frontier was the lawless tribal area between British India and Afghanistan. In this area, the British government was represented by Frontier Agents. These were Englishmen, but they were also men who had lived in the area for a long time and knew the local players and politics well. Their actual power was trivial, little more than some cash and a company of Indian troops called Sepoys. But that small power was often enough to tilt the local political and military balance for or against a local chieftain.

The local leaders were aware of this, and they usually found it worth their while to maintain good relations with the British so as to keep them on their side, or at least keep them from actively intervening against them. Here the key is good local intelligence, especially political intelligence. By integrating with the local population, state forces can learn the existing local political divisions and alignments in order to utilize them. As the Northwest Frontier Agents once did, they can leverage local relationships to achieve their ends while avoiding unnecessary combat.

The Primary Fourth Generation Supporting Arm: Cash

What artillery and air power are in Third Generation war, cash is in the Fourth Generation: the infantry's most useful supporting arm. Local commanders must have a bottomless "slush fund" of cash to use at their discretion. Obviously, this cash cannot be subject to normal accounting procedures as most will, necessarily and properly, be used for bribes. It is imperative that any regulations that present legal or bureaucratic obstacles to this bribery are promptly changed.

One way to do this might be to establish the billet of "Combat Contracting Officer." The Combat Contracting Officer would have legal authority to pay money as he sees fit in order to support the commander's objectives. This would include payments to get local services operating quickly, support local political leaders who are working with state forces, and obtain local resources. It would explicitly include the authority to pay bribes. That is how much of the world works, and if foreign forces are to obtain results they must be able to adjust to the world in which they find themselves rather than expecting the world to operate as they think it ideally should.

The Fourth Generation's Geneva Conventions: Codes of Honor

While state armed forces will remain bound by the Geneva Conventions in Fourth Generation conflicts, their opponents will not be. Non-state forces are not party to inter-state international law. However, in some cases it may be possible to agree with Fourth Generation opponents on a chivalric code of sorts that sets rules both sides are willing to follow. Some Fourth Generation entities have self-images that make honor, generosity, and lineage tracing to noble forebears important to them. Just as chivalry was important before the state, it may again become important after the state. Where these attributes are present, it may be to state forces' advantage to propose a Fourth Generation code of honor.

The specifics of such a code would vary from place to place. It might include provisions such as the state forces agreeing not to use air bombardment while their opponents agree not to set off bombs in areas where civilians are likely to be present. Regardless of the specifics, the use of such codes will generally work to the state's advantage. They will diminish the counterproductive "Goliath" image, demonstrate that state forces respect the local people and their culture, and generally help de-escalate the conflict. They will also assist in improving public order, which in turn helps in preserving or recreating a state. Disadvantages such codes may bring to state forces at the physical level will generally be more than compensated by advantages they bring at the mental and moral levels.

The "Mafia Model": Everyone Gets a Cut

Just as the Northwest Frontier Agent offers us some useful ideas for Fourth Generation conflicts, so does the "Mafia Model." Various mafias have run Sicily and southern Italy for decades, if not centuries. How has it succeeded in keeping the peace? How would the mob do an occupation?

One key to a mafia's success is the concealed use of force. If an individual needs to be "whacked," then it is usually done with little fanfare and in the shadows. The rule is, "No fingerprints." Unless there is a specific message intended for a larger audience, people who are killed by the Mafia are seldom found.

This method usually requires patience. It often takes a long time for the right situation to present itself. If there is a message to be made to a larger audience, then a public display of violence can be used. But this is often best avoided, as it can backfire against the aims and goals of the organization due to the negative effect on public opinion it often entails.

The Mafia also operates on the principle that "everyone gets his cut." If you are willing to work with the Mafia, you get a share of the profits. Money is a powerful motivator, especially in the poorer parts of the world where most Fourth Generation conflicts occur. In working with the local population, state forces should carefully design their approach so that everyone who cooperates with them gets a financial reward. The rewards should grow as the "business" expands, that is, as state forces get closer to achieving their objectives. This is especially important for leaving a stable situation behind when foreign forces withdraw. If everyone is profiting from the new situation the foreigners have created, they will be less eager to overturn it and return to instability.

Thoughts on Fourth Generation War Techniques

Third Generation militaries recognize that any technique usually has a short shelf life in combat. As soon as the enemy comes to expect it, he turns it against you. This, in turn, means that the ability to invent new techniques is highly important. Units that develop a successful new technique should communicate their discovery laterally to other units. Fourth Generation war makes this all the more important, because

Fourth Generation opponents will often use techniques very different from those used by state-armed forces.

Here are a few techniques for Fourth Generation war, provided as examples. The purpose of providing them is to illustrate the creative and very different way of thinking that will be required to develop effective techniques for Fourth Generation conflicts.

- Equip every patrol with a camera. If the patrol is fired on, it attempts to get a picture of those doing the firing. Then, contracts can be put out on those who can be identified.

- Sponsor a local television program where captured enemies who have killed civilians are interrogated by the local police. This was highly effective in Iraq.

- Distinguish between captured opponents on the basis of motivation, tribe, religion, or some other basis that local people will recognize. Treat some as "honored guests" and send them home, while continuing to detain others. This can cause suspicions and divisions among opponents.

Intelligence in Fourth Generation Warfare

The current military intelligence model is antiquated philosophically, structurally, and procedurally. Philosophically, it assumes that higher headquarters have a clearer picture of the situation, which they provide to tactical units. Structurally, it concentrates resources, especially in the form of trained intelligence analysts, at higher command levels. Procedurally, it follows (at least in the U.S. military) a system known as Intelligence Preparation of the Battlespace (IPB) that was developed during the Cold War and focuses on counting and "templating" major enemy units.

This legacy model is not appropriate for Fourth Generation war. The granular nature of Fourth Generation battlefields means that each company may face a different situation, which it knows much better than any higher headquarters can. Intelligence must be a bottom-up, not a top-down process; higher-level headquarters will create an intelligence picture largely by piecing together the analyses of small units.

This in turn suggests that intelligence assets, especially trained personnel, should be pushed downward, to the battalion and company levels. At present, those levels gather far more intelligence than they can process. Local processing of intelligence reduces the distortion that inevitably accompanies transmission to higher headquarters. It also moderates the demands for information that higher headquarters place on small unit commanders, which have reached dysfunctional proportions.

In turn, IPB is quantitative in nature and must be replaced by qualitative assessments of the enemy and the local population. Instead of training intelligence personnel in a rote method, we should educate them broadly so they can develop an instinctual feel for the situation, including its historical, cultural, and ethnic components. Only then can they provide commanders a comprehensive intelligence orientation that can serve as a basis for clear decisions and effective actions.

The Swedish approach to intelligence may provide a useful model for Fourth Generation conflicts. The Swedish word for military intelligence is *Underraettelser*. The term is a combination of two words, *under* and *raettelse*. *Under* means "from below" and *raettelse* means "correction." The word translates literally as "corrections from below." Every Swedish Marine is an intelligence collector. The Swedish company commander trusts his Marines to give him the intelligence he needs and the battalion commander trusts his company commanders to tell him what he needs to know. He may then activate his reconnaissance units or special forces to focus on specific questions of direct interest at his level.

Knowing that the best intelligence comes from the lowest level, dissemination of that intelligence must be allowed at the lowest level. The right to spread information and to act upon it must be decentralized. Decentralizing the right to coordinate intelligence across organizational lines and facilitating the ability to act upon it is not a new feature of Fourth Generation war, but it is even more important in Fourth Generation war than in Third Generation war.

Fourth Generation War and the Press

State militaries can take two different general approaches to the press, defensive or offensive. In the defensive approach, the objective is to minimize bad press by controlling the flow of news. This was typical of how militaries approached the press in Second and Third Generation wars. The offensive approach seeks to use the press more than to control it, though some control measures may still be in place. Many Fourth Generation entities are highly effective in using the press, including the informal Internet press, for their own ends. If state-armed forces do not also undertake a press offensive, they are likely to find themselves ceding to the enemy a battlefield that is important at the mental and moral levels.

In turn, the key to an offensive press strategy is openness. Few members of the press or media such as the Internet will allow themselves to be so controlled as to present only the good news. Unless state forces are open about mistakes and failures, the press will devote most of their effort to ferreting them out. Worse, state forces will lack credibility when they have real good news to present.

Paradoxically, openness is the key to controlling negative information in the few situations where that is really necessary. Sometimes, openness builds such a cooperative relationship with the media that they become part of your team and do not want to report something that will really hurt you. At other times, you can use the credibility you have built through a general policy of openness to deceive when deception is absolutely necessary. Just remember that when you do so, you may be spending your only silver bullet.

Winning at the Mental and Moral Levels

At the mental level, Fourth Generation war turns Clausewitz on his head. Clausewitz wrote that war is the extension of politics by other means. At the mental level of Fourth Generation war, politics is the extension of war by other means. Not only are all politics local, but everything local is politics.

To win, state forces must learn how to make the local politics work toward the ends they are seeking. If they fail, no military gains will last once they depart, as at some point they must, at least in the case of foreign forces. Much of this handbook has been devoted to what state forces must do to succeed in the local political environment, including understanding the local culture, integrating with the local population, and developing an effective bottom-up intelligence system.

At the most powerful level of war, the moral level, the key to victory is to convince the local people to identify with the state, or at least to acquiesce to it, rather than identifying with non-state entities. Meeting this challenge will depend to a significant degree not on what state forces do, but on what they do not do. They cannot insult and brutalize the local population and simultaneously convince them to identify with the state. They cannot represent a threat to the local culture, religion, or way of life. They cannot come across as Goliath, because no one identifies with Goliath. Nor should they come across as Paris, the Trojan champion in the *Iliad*, who, as an archer, fought from a distance and was therefore considered a coward.

This does not mean state forces should be weak, or project an image of weakness. That is also fatal, because in most cultures, men do not identify with the weak. History is seldom determined by majorities. It is often determined by minorities who are willing to fight. In most Fourth Generation situations, the critical constituency the state must convince to identify with or acquiesce to it is the young men of fighting age. To them, state forces must appear to be strong without offering a challenge to fight that their honor requires them to accept. They may identify with an outsider who is strong, but they will fight any outsider who humiliates them.

In terms of ordinary, day-to-day actions, there is a Golden Rule for winning at the moral level, and it is this: Do not do anything to someone else that, if it were done to you, would make you fight. If you find yourself wondering whether an action will lead more of the local people to fight you, ask yourself if you would fight if someone did the same thing to you. This Golden Rule has a corollary: When you make a mistake and hurt or kill someone you shouldn't or damage or destroy something you shouldn't—and you will—apologize and pay up fast. Repair and rebuild quickly, if you can, but never promise to repair or rebuild and then not follow through.

This brings us to the bottom line for winning at the moral level: Your words and your actions must be consistent. We deliberately have not talked about Psychological Operations (PsyOps) in this handbook, because in Fourth Generation war, everything you do is a PsyOp—whether you want it to be or not. No matter what the local population hears you say, they will decide whether to identify with you, acquiesce to you, or fight you depending on what you do. Any inconsistency between what you say and what you do creates gaps your enemies will be quick to exploit.

Chapter 3

The Fourth Generation Warfare Counterforce: Light Infantry

The purpose of this section of the handbook is to describe the nature, tactics, and training of light infantry, which in most cases is the best force to employ against Fourth Generation opponents. The section has been organized in three chapters. This chapter defines what light infantry is. Chapter 4 explains how light infantry fights. Chapter 5 discusses how to convert line infantry into light infantry, i.e. how to train.

The History of Light Infantry

Due to different meanings of the word "light," light infantry has been understood in diverse ways around the world. These interpretations can be grouped into two different points of view. The present American concept of light infantry is related to weight, specifically weight of equipment, while Europeans understand "light" as relating to agility or operational versatility. They see light infantry as a flexible force capable of operating in austere conditions with few logistical requirements and employing tactics unlike those of line or mechanized infantry.

The distinction between regular or line infantry and light infantry goes back to ancient Greece. At that time, the regular infantry was the phalanx, a linear formation that based its power on mass and shock. Their tactics consisted of evolutions performed by the phalanx as a whole, in which each warrior adhered to carefully executed drills.

In contrast, classic light infantry did not fight in fixed formations, nor did it adhere to any type of prescribed methods. Its primary mission was to provide flank protection to the phalanx. Widely dispersed throughout a large area, its soldiers lacked the heavy bronze armor worn by hoplites. The survivability of the light infantry depended on speed and the use of bows, slings, and hand-thrown weapons. Light infantry tactics consisted mainly of individual actions or simple, loosely coordinated group maneuvers that were generally limited to advancing or withdrawing. The Romans applied the Greek concept to their legions, using light auxiliary infantry to support the heavily armored cohorts of their regular infantry.

After the medieval era, when cavalry ruled the battlefield, the Spanish *tercios* of the 16th and 17th centuries signaled the return of the infantry's dominance. The development of light infantry in Europe followed in the 18th century. The French *Chasseurs*, the Prussian *Jaegers*, and the Austrian *Grenzer* regiments followed the ancient Greek concept; in contrast to the rigid maneuvers of their line infantries, the light units were fast, agile, and expected to adapt their tactics to the terrain and the situation.

Much as their predecessors had been in the past, the Napoleonic light infantry was employed in a decentralized manner to protect the flanks of larger forces and to execute raids and ambushes in restricted terrain. As before, the light infantry was always careful to avoid frontal engagements with the enemy. When it was wisely employed, light infantry could sometimes prevail over the enemy's regular infantry thanks to its adaptability and reliance on creative tactics rather than drilled battlefield order. These capabilities were achieved by selecting high-quality troops to serve in the light infantry, often professional hunters or foresters.

In spite of the proven utility of light infantry units, they were not established as permanent formations in European militaries. Light infantry units only prospered during wartime, and they were usually dissolved when the conflict ended. The catastrophic defeat in 1755 in Pennsylvania of the British forces under General Edward Braddock by a small force of Indians and French light infantry that employed ambush tactics and took advantage of terrain, agility, and loose formations convinced the British to create Roger's Rangers and the Royal American Regiment, both of which eventually became famous light infantry units during the French and Indian War. Typically, both were dissolved when the war ended.

Light infantry reappeared in Europe during the wars surrounding the French Revolution. The light infantry ceased to be regarded as an "undisciplined group of irregulars" and were transformed into trained professional units, able to maneuver in a decentralized but fast and organized manner. Between 1790 and 1815, light forces proliferated, even evolving into light artillery and light cavalry units. They also assumed a more significant role on the battlefield. Yet their basic role remained no different than that of their ancient Greek predecessors, as the European light infantrymen covered the regular infantry's advances and withdrawals and harassed the enemy by executing ambushes deep in their rear.

The appearance of the breech-loading rifle and the machine gun gradually reshaped regular infantry tactics, which began to resemble more closely those of light infantry. However, true light infantry retained advantages in agility, operational versatility, capability for living off the land, and decentralized command and control. The Boers of the Transvaal Republic; the Jaeger battalions, mountain units and *Sturmtruppen* of the German army of World War I; General Wingate's Chindits; and the paratroop units of the Israeli Defense Forces and the British army are examples of true modern light infantry.

The Light Infantry Mentality

The appearance of semi-automatic and automatic weapons narrowed the tactical differences between light infantry and regular infantry. However, the essential difference between them remains. It is not easily observed because it is an intangible factor: the *mentality* of the light infantrymen.

The light infantryman is characterized by his mental resourcefulness and physical toughness. Light infantry's inborn self-reliance, reinforced by hard training, convinces the light infantryman that he is capable of overcoming the most difficult situations that combat presents. Light infantrymen do not feel defeated when surrounded, isolated or confronted by superior forces. They are able to continue performing their duties and pursue their objectives for long periods of time without any type of comfort or logistical support, usually obtaining what they require from the land or the enemy. They are neither physically nor psychologically tied to the rear by a need to maintain open lines of communication. Their tactics do not depend on supporting arms. This attitude of self-confidence and self-reliance provides light infantry a psychological advantage over its opponents.

Thanks to its decentralized command philosophy, light infantry operates at a high tempo. An ambush mentality, a preference for unpredictability and a reluctance to follow rigidly specified methods are the essence of light infantry tactics. The ambush mentality generates other secondary light infantry characteristics. One is the speed with which light infantry adapts to the terrain. Far from resisting adverse environmental conditions, light infantry exploits them by turning rough terrain to its advantage, using the terrain as a shield, a weapon, and a source of supplies.

As a result, light infantry has an incomparable superiority in those terrains that restrict most regular infantry operations (especially mechanized forces), usually allowing the light infantry to face and defeat larger and better-equipped enemy forces whenever it encounters them. This advantage gives the light infantry its distinctive operational versatility, as it is able to operate alone in restricted terrain or in a symbiotic relationship with friendly units.

Light infantry is readily adaptable to a broad range of missions, and it faces the natural evolution of technology and tactics that always takes place in wartime with no need to substantially modify the way it operates. It should now be easy to see that the correct meaning for the term "light" is not the American notion of weight, but the European concept of agility and operational versatility.

Chapter 4

Light Infantry Tactics

Light infantry tactics are offensive in character, even during defensive operations. Light infantrymen do not hold a line. Light infantry tactics follow the principles of maneuver warfare, attacking by infiltration and defending by ambush. It uses ambushes on the offensive as well, by ambushing withdrawing or reinforcing enemy units, sometimes deep in the enemy's rear. Light infantry applies an ambush mentality to both planning and execution.

A good way to understand light infantry tactics is to think of them as similar to those often used by "aggressors" or the "red team" during training exercises. Lacking the means to execute their missions in textbook fashion, they fight by deceiving, stalking, infiltrating, dispersing, looking for vulnerabilities, ambushing, and raiding. They often prove highly effective against larger "blue" forces.

Light infantry operations often follow a cycle that can be divided into four steps: Dispersion, Orientation, Concentration, and Action (DOCA). Dispersion provides light infantry with its main tool for survivability. Units remain hidden, taking advantage of the terrain, using camouflage and fieldcraft to evade detection. Orientation includes shaping actions that "set up" the enemy and permit rapid concentration. This step requires an aggressive use of reconnaissance to identify enemy vulnerabilities the light infantry can exploit.

Concentration allows light infantry to transform the small combat power of many dispersed elements into one or more powerful thrusts. Action is led by reconnaissance elements, which focus available forces and target a specified enemy weakness. Finally, a new and rapid dispersion ends the cycle, protecting the light infantry from enemy counteraction.

Light infantry offensive tactics usually use infiltration to avoid casualties. Infiltration allows light forces to surprise the enemy and engage him at short distances. In close, light infantry can exploit its small arms skills while denying the enemy effective employment of his superior firepower. Light infantry hugs the enemy and forces him to fight at short ranges on its terms.

Defensive Tactics and "Force Protection"

Light infantry defenses are dispersed and granular, which prevents the enemy from determining the exact location of the defense's front, flank, or rear areas. This protects light infantry from concentrated firepower. The light infantry commander assigns sectors to each of his subordinates, areas where they plan and execute successive, independent ambushes on advancing enemy formations. The "baited ambush" is a common technique, where a unit will feign retreat or even rout to draw enemy units into a new ambush. Defenses run parallel to, not across, enemy thrust lines. Light infantry often focuses its efforts against follow-on enemy units rather than spearheads.

When threatened, light infantry units break contact and move to alternate positions, setting up a new array of interconnected ambushes. Light infantry never fights a defensive battle from fixed positions or strong points. From the light infantry perspective, a good defensive position is one that surprises the enemy from a short distance, but at the same time enables the defender to move fast and under cover to a new position unknown to the enemy.

Since light infantry lives mostly off the land, its success depends heavily on the support of the local population. This dependence on local support means light infantry operations always need to avoid a negative impact on the inhabitants and the local economy, as well as rigorously observe local customs and culture. This ties in directly with requirements for success in Fourth Generation wars.

Light Infantry vs. Fourth Generation Opponents

Most Fourth Generation forces are light infantry, some quite good, for example, Hezbollah and the Pashtuns. How does state light infantry defeat them? By being *better* light infantry than they are.

Fourth Generation war light infantry is likely to have some advantages over state light infantry. It will usually know the terrain better. It is likely to start out with stronger support among the local population, especially if the state forces on the other side are foreign. That support will mean a superior information network, among other benefits.

But at the tactical level, state light infantry should usually be the more skillful force. State light infantrymen are full-time soldiers, while most Fourth Generation fighters will be part-time militiamen. State forces have more resources for training, better equipment, better logistics, and sometimes in combat they can employ supporting arms, which they use when they can although they do not depend on them. State light infantry should be more skilled at techniques, including marksmanship and tactical employment of machine guns and mortars. Assuming they can at least match their Fourth Generation enemies in tactical creativity, their superiority in techniques should often be decisive.

The superiority of state light infantry does depend on their being employed correctly. If they are compelled to defend static positions, given detailed, controlling orders, overburdened with weight (they should seldom if ever wear body armor or helmets; the soldier's load should not exceed 45 pounds), or tied to supporting arms or to communications "networks" that require constant input, they will lose the advantage they should have over non-state light infantry. Requiring cats to hunt like dogs will benefit only the mice.

Chapter 5

Light Infantry Training Objectives

Before talking about how to train light infantry, and specifically how to convert line infantry to light infantry, we should know what qualities our training should inculcate. They include:

Patience. The need for patience is, perhaps, the greatest difference between light infantry and line infantry. Light infantry operations proceed much more slowly, primarily due to the requirement for light infantry to operate stealthily. It takes time to discover targets, reconnoiter suitable ambush sites, and move covertly. Training must reflect this. The tempo of operations will slow down and light infantry training schedules must come to grips with this fact. One method is to schedule open-ended training in which the exercise does not end at a predetermined time; it ends only when the training goals are accomplished.

Speed. While setting up a light infantry action requires patience, when action occurs, it must be over fast, before the enemy can react. The light infantry then normally goes covert again. Decision-making in the light infantry is also characterized by speed. Light infantry leaders must be prepared to react immediately to unforeseen situations with changes in their plans. They seldom have the luxury of other forces coming to their rescue. They cannot afford to be pinned down physically or mentally. Snap decisions often mean the difference between success and failure. Delaying a decision once action commences is usually dangerous.

Self-discipline. Self-discipline may be the most important quality in light infantrymen. Without self-discipline, individuals will be unable to cope with the privations and stresses that are an inherent part of being light infantry. Troops that do not demonstrate self-discipline are a positive danger to the mission and to their comrades and have no place in the light infantry.

Therefore, one of the most important goals of training light infantry is to substitute self-discipline for imposed discipline. Senseless or unnecessary rules should be done away with. As maneuver warfare doctrine requires, orders should normally specify only the result to be obtained, not methods. Leaders should expect and encourage their troops to maintain a high level of discipline, not through fear of reprisal, but because of a desire to demonstrate their professional qualities.

Physical fitness. Being a light infantryman is physically demanding. Troops must maintain the ability to move long distances quickly, with or without loads. At the same time, the "soldier's load" should not exceed 45 pounds, beyond which march performance is degraded regardless of physical conditioning.

Light infantry also requires a different approach to physical fitness from that currently taken by many state militaries. Light infantry requires endurance far more than physical strength. Soldiers must be able to march long distances; they must be prepared to move all night carrying their combat load in difficult terrain. Physical fitness standards for light infantrymen should reflect this emphasis on endurance.

"Jaeger" mindset. Light infantrymen are hunters on the battlefield and every effort should be made to impress this upon new members of the unit. All hunters require fieldcraft of a high order. Light infantry should hunt enemies the same way they hunt game.

Stalking skills. Good stalking skills are essential in order to surprise and ambush enemy forces. Poor stalking skills expose a light infantry force to detection, which often means defeat and destruction.

Proficient with organic and threat weapons. Every soldier must be intimately familiar with all of the weapons found throughout his unit. This is particularly important in light infantry units which operate primarily as small units; it provides a high degree of flexibility to the unit, especially when it suffers casualties. The ability to utilize threat weapons allows light infantrymen to use captured items, which may at times be all that is available. Light infantry units in combat have only occasional, not continuous, logistics pipelines.

Comfortable operating at night and in varying terrain. Because stealth is so critical to the survival of light infantry, it is important that light infantrymen are able to operate effectively at night. In fact, the vast majority of light infantry training should be focused on improving proficiency operating at night. It is impossible to know where the next conflict will occur, so light infantry must also be able to operate in any type of terrain, except open terrain, where all foot-mobile infantry is vulnerable to being bypassed and pocketed by mechanized forces.

Proficient utilizing demolitions. Given the fact that light infantry operates most effectively in small units, every light infantryman should be capable of utilizing demolitions. Demolitions are an important element in ambushes or raids, staples of light infantry offensive actions. Light infantrymen know how to improvise explosives where necessary. One writer on tactics noted, "An illustrative difference between light and line infantry is how each uses chicken shit. Light infantry uses it to build IEDs [improvised explosive devices]. Line infantry employs chicken shit to consume its own time."

Broad perspective. Light infantry must consider their actions in the widest context possible. In referring to the actions of light infantry more than 200 years ago, Johann von Ewald, a Hessian Jaeger company commander, wrote that a light infantry leader "has to do on a small scale what a general does on a large scale." This means that light infantry leaders must consider how their actions will impact the mission at the highest levels. While an action may be beneficial at a local, tactical level, will it aid in achieving victory at the more powerful operational or strategic levels, or the mental and moral levels? All leaders, no matter how junior, must be educated, encouraged and rewarded for thinking in as broad a context as possible.

Chapter 6

Operation Pandur

Nobody wanted to say it, but the Marines of K.u.K. Marine Battalion 3, Company 2 knew that they had gotten their butts kicked—again. This time, two Marines were dead, seven had to be MEDEVACed and the mission had to be cancelled. It wasn't the first time. Sometimes, the only thing that saved the company from being overrun was support from Marine air.

By all normal accounting, the Marines knew they should win every engagement. Their enemy had none of their advantages. He didn't have any air support or air reconnaissance. He had to leave his badly wounded on the field for Marines to take care of, because he had no medical transport. None of his men had been to boot camp, the School of Infantry (SOI) or The Imperial Basic School (TIBS).

He was good at only two things. He knew how to place improvised explosive devices, and he knew how to appear out of nowhere, ambush 2nd Company, then vanish. But at those two things, he was very, very good. The enlisted Marines called their enemies "ghosts."

After they got back to base, just as the company commander, Captain Trapp, had taken off his gear, the company gunny came up to him. "Sir, may I speak frankly?" asked Gunny Blau.

"Of course, Gunny," replied Captain Trapp. Trapp knew his Marines and he had a pretty good idea what was coming.

"Sir, the men have had it. We're tired, morale's hanging down lower than a Bassett's balls and we're effing sick of getting beat. There has to be a better way to fight this goddamn war than wandering around waiting to get blown up or ambushed by someone who vanishes into thin air before we can hit him back. Sir, you're our leader. Isn't there something we can do differently?"

Captain Trapp sat down on his pack. He had been thinking the same thing for a long while now. The time had come to decide and act.

"Yes, Gunny, there is. We can become what our enemy is. We can become light infantry," he said.

"I thought we *were* light infantry," the Gunny replied. "We're not mech infantry."

"I know that is what you were taught, Gunny, but that isn't the main difference," Trapp said. "True light infantry has a whole way of fighting and thinking that is different than what we do. We're line infantry, not light infantry, and that is the core of the problem. And I know what we have to do about it."

When Captain Trapp went through TIBS, the TIBS Commanding Officer had been a colonel—one of the few—who knew the difference between line and light infantry. He had personally taught an after-hours seminar for those lieutenants who were interested (most were not) in true light infantry tactics. Trapp took the seminar. He read things like Johann von Ewald's *Treatise on Partisan Warfare*, Franz Uhle-Wettler's *Battlefield Central Europe: The Danger of Overreliance on Technology*, and Steven Canby's *Classic Light Infantry and New Technology*.

The next morning, Captain Trapp went to see the battalion commander. He found Lieutenant Colonel Franz Josef von Stahremberg outside chewing on a cigar and swearing quietly to himself.

"How are you this morning, sir?" asked Trapp.

"Pissed off at the world, Captain. I'm frustrated. The firefight you were in yesterday seems like every other fight each of the battalion's companies get into. I'm tired of our troops chasing ghosts and catching crap. When we do find the enemy, it's because he wants to be found and is waiting for us. Then, he vanishes either into the terrain or into the civilian population. There's got to be a better way to do this!"

"That's exactly what I wanted to talk to you about, sir. I feel the same way, and so do my men," said Trapp. "The problem is that we're not really light infantry, sir, but the enemy is. What we need to do is beat the enemy at his own game."

"What do you have in mind?" asked the lieutenant colonel.

Trapp decided to dive right in, "Take my company off-line, sir. Let me retrain them as true light infantry. I stayed up last night working on a training plan. Give me a month and I'll be able to retrain my Marines to operate as light infantry at the squad level. In another month, we'll be proficient at the platoon level and in three months, we'll be capable of company operations."

Von Stahremberg took a moment to think. "You're asking a lot, Trapp. As it is, I don't have the troops to cover our entire area of operations and you want me to take your entire company off-line? How can I do that?"

Trapp was ready with his reply, "If we don't retrain our Marines, sir, then we can't expect things to change. We're both tired of Marines getting killed without the ability to hit back. Becoming proper light infantry will give us the chance to hunt the enemy rather than being hunted."

There was another pause as von Stahremberg considered his options. "All right, do it," he said. "I'm ready to try anything that will improve our chances against these guys. I'll talk to the Operations Officer and Sergeant Major and let them know what's going on."

Trapp immediately returned to his company headquarters to put into action the training schedule he had developed. The first thing was to inculcate in his Marines a light infantry mindset. They would have to be prepared for anything at any time, day or night, and they would have to be able to improvise if they did not have the proper gear or some drill learned at a school. Developing this mindset would be a theme that ran throughout the training. Trapp decided that he would not publish his training schedule. At all costs, he wanted to keep his Marines from becoming comfortable or complacent.

Trapp knew he needed to explain to his Marines what they were doing and why. The troops had to know what it meant to be light infantry and how that was different from what they had been taught in their training. He ordered the Company First Sergeant to get the Marines together.

Captain Trapp knew he would have to undo in weeks what the K.u.K. Marine Corps had spent years in creating. Undoubtedly, some of his Marines would not be able to measure up. Those that did not possess minds agile enough to adapt quickly or that could do little beyond implementing a checklist would likely have to go. He would have to be ruthless in his evaluations.

As Trapp stepped outside the tent, his head cleared. Stepping in front of his men, he felt sure of what he was doing.

"Marines, we've had a tough couple of days. In fact, we've had a tough deployment so far. What we've been doing hasn't been working, and we all know it. So we're going to change. We're going to become a real light infantry. Some of you might think we're already light infantry. That's not correct. What we're going to do is learn how to beat the enemy at his own game. We've got some ground to make up, but we're going to be better because we're going to train harder than he does. He's got a day job to keep up. This is our day job.

"What does it mean to be light infantry? It means you are a hunter. How many of you hunt?"

Most hands went up.

"Would you hunt a deer the way we've been operating?"

"Hell, no!" came a reply from the back of the group.

"Well, we're going to start hunting the enemy the way you would hunt a deer. From now on, you live out of your pack and your pack will be light. You will stalk the enemy instead of letting him stalk you. You will go where he thinks you can't and ambush him where he thinks he's safe. You will learn to move without being seen, ambush the enemy and disappear. If he's a ghost, you will be wraiths.

"From now on, you will live off the land, sleeping outside with no Internet and no contractor-run chow halls. You will welcome self-discipline, hard work, and being ready for anything at all times.

"Put together, all this means that we will finally have the advantage when we take on the enemy. He will learn to be afraid of us because he will never know where we might hit him or how. Starting now, we will learn to be better light infantry than he is."

The Marines were listening intently as Trapp spoke. He could see they were engaged. Most of them had lost good friends and all of them were eager to give the enemy a taste of his own medicine. But he could also see uncertainty in some faces. He was asking them to take a leap of faith and forget much of what they had been taught. He paused for a moment, knowing that any doubts would grow with his next order.

"I want all these tents folded up and returned to supply. Sort all of your gear and personal effects. If you keep it, you will have to carry it. Your total load may not exceed 45 pounds, and I will weigh it. Store whatever you are not going to carry in your seabag. All seabags will be turned in to supply. We sleep wherever we are, under the stars, rain or shine. You have two hours to get this done. Move."

Trapp turned away as pandemonium broke out in his wake. No one knew that in just one hour, he would order the company to hike the short distance to the camp's small arms range. That afternoon, while the company conducted a small arms shoot with the marksmanship of all hands individually evaluated, the company area would be swept clean. All remaining seabags would be taken to supply. Tents left in various states of disorder would be rolled up and taken as well. When the company returned from the range, they would find no trace of their previous existence. The Marines would be unencumbered and free to go wherever their missions took them.

Working on the company's mindset as light infantrymen was the first step. Trapp's next was to make all of his Marines proficient with all of the unit's weapons. Every Marine had to know how to employ machine guns, which would be among the most important weapons in any ambush. Marines also needed to know how to fire the 60mm mortar in the hand-held mode. Trapp's Marines moved from station to station, learning to disassemble, assemble, and maintain the weapons when they were not on the firing line actually shooting them. Trapp wanted to ensure that all of his Marines (even his corpsmen) were proficient with each weapon they might be called upon to use in combat.

Trapp next had to train the Marines to use the enemy's weapons. While the battalion had not taken many weapons in combat with the enemy, they had found a number of caches. These caches provided all of the weapons needed to teach Trapp's Marines how to shoot, break down, clean, and reassemble everything the enemy used except IEDs. The enemy's ammunition was not stored or cared for very well, so Trapp asked the Battalion Gunner to teach his Marines how to inspect and clean captured ammunition to ensure it was safe to use. After a week and a half of non-stop live fire, including becoming comfortable using grenades at close quarters, Trapp was confident that his Marines could use any of his unit's weapons, or any of the enemy's, that came to hand.

One of the most important tenets of all of Trapp's training was that it had to be truly free-play force-on-force training. In every training event he planned, there had to be an untethered opposing will. In the past, he had heard his superiors go on endlessly about the training standards for a given action, for instance, how much time was acceptable to set up mortars for a hip shoot. He had found that none of this made any difference in combat. The only thing that mattered was whether or not his troops were better than the enemy, could out-fight the enemy and, most important of all, could out-think the enemy. The ability to do these things could not be created, measured, or improved by any comparison against irrelevant "standards." Trapp had found that qualitative factors, not quantitative factors, were more important in determining a unit's quality and capabilities. In Trapp's mind, the only true measure, the only measure that really mattered, was how his force measured up against his enemy's hostile, independent will.

Trapp knew that free-play training not only provided the best training for his men, but also contributed to the creation of new techniques that could be shared throughout his unit. To further facilitate his men's creativity, he decided to issue intentionally vague orders in training. He would tell each side to "destroy the enemy force" and let them figure out how to accomplish this goal. To remind his troops of the need to discard any preconceived notions regarding what they must or must not do, Trapp also decided the company would adopt two mottos. The first would be, "Do what works." The second would remind them of the freedom they had as aggressors in other training exercises, "Every Marine an aggressor *all the time.*"

As he considered his training plan, Trapp decided that it was best to take a building-block approach and start with the skills he wanted every Marine to possess. Once those were cemented, he could move on to the squad level, then to the platoon level and finally to the company level. Trapp ticked off the list of training goals that he still needed to accomplish: fieldcraft, especially concealed, soundless movement; basic survival skills to allow his Marines to live off the land for short periods of time; demolitions to destroy enemy equipment that could not be carried away and also to initiate ambushes; and advanced life-saving skills to keep Marines alive who could not immediately be evacuated. Last would come supporting arms training. Supporting arms were useful, but he wanted to break his Marines' dependence on external agencies. They had to become self-reliant.

In the confines of the forward operating base, Trapp's Marines became an object of intense interest to other units. They were firing weapons of every description day and night. When they were not shooting, they walked for hours along the circular perimeter berm, with their crazy company commander occasionally preparing surprises for them. On one occasion, he quickly designated one Marine in each squad a casualty and forced the squad to build a stretcher out of whatever they had with them and to carry the individual for the next hour. During one march, squad leaders were called away to conduct a sand-table exercise (sand was one thing the Marines had plenty of) and while their leaders were away the squads were "ambushed" by host-nation militiamen.

Captain Trapp carefully observed his leaderless squads' reactions. Some reacted quickly without orders, others hesitated. Sometimes, the senior remaining Marine in the squad led the group. In other cases, it was a junior Marine. Trapp took note of the informal leaders who showed promise and quietly spoke to the Company First Sergeant about them. They would be given their stripes if they continued to develop.

One day the battalion commander approached Captain Trapp. He had a serious look on his face.

"Johann, we've been holding the fort with a skeleton force while you trained. I've given you a few weeks. Are your Marines ready to operate as squads yet? We need help out there."

Trapp was uncertain. He wished he could complete the entire training program before he committed his Marines to combat. He wished he could have conducted all of this training back home, before the deployment. He knew that the pressure on the battalion was mounting, but he didn't want to send his Marines out just to become targets again if they were not ready.

"Sir, about three-quarters of my squads are ready. The others are not too far behind. If we rotate the Marines outside the wire, so that about half are out at any given time, that will allow me to conduct concurrent training with the Marines that are at the forward operating base."

"How long will your squads stay out?" asked Lieutenant Colonel von Stahremberg.

"That depends on their mission, sir. They can stay out for a while without being resupplied. They can buy food in a few places, and in the higher areas, they can trap some game. They won't eat well, but they won't starve. The bigger concern is water. There are not many water sources, so they'll probably have to sneak it from someone's well in the dark of night. The mission will determine how long they stay out, rather than logistical considerations. The other concern I have, sir, is that we haven't had the chance to work on the Marines' land navigation skills because we're confined to the FOB. In the initial stages, I don't want to push the Marines too far. As they become more proficient navigating at night with a map and compass, not just GPS, we can push them further out.

"I've been working with the Intelligence Officer on some missions that I think will meet the battalion's needs as well as suit the limitations I've just described, sir."

"What have you got?" asked von Stahremberg.

In the discussion that followed, Trapp outlined his plan for the battalion commander. Squads would be inserted by a variety of covert methods over a 24-hour period. Each squad was to make its way to a different enemy-controlled area and establish ambushes. Some squads would place ambushes near suspected enemy infiltration routes. Others would establish ambushes along the routes the battalion usually used.

The Marines' logistics convoys had been ambushed repeatedly by the enemy from high, rocky terrain at a bend in the main supply route. The Marines had never been able to catch the enemy and had tried everything from ground sensors to preemptive artillery strikes in an effort to disrupt the insurgents. Nothing worked for long. Trapp believed that the insurgents came from one of the villages to the north of this hill mass. They usually ambushed the convoys with RPGs and medium machine guns. The fact that the insurgents conducted attacks in the same general area and the speed of their withdrawal afterward indicated that they probably cached weapons and ammunition in the vicinity of their ambush site.

Trapp planned on accompanying the squad tasked to establish this counter-ambush. This would allow him to demonstrate to his Marines that he asked nothing of them that he was not prepared to do himself. He also wanted to see how the squad leader conducted himself. The Company Executive Officer could hold down the fort back at the FOB this time. There should not be much for him to do other than conducting the training that Trapp had planned. The squads that were hunting the enemy would operate under radio silence.

The next night, Trapp found himself rolling out of the back of a slowly moving local civilian truck. He and his Marines had been concealed for nearly an hour as part of a convoy moving in the dead of night. The Marines moved quickly away from the road and established a hasty perimeter, waiting in case the enemy had discovered them and allowing their senses to adjust from the noise of the trucks to the quiet that now enveloped them. Moving closer to the squad leader, Lance Corporal Hummel, Trapp was gratified to see that Hummel had carefully tracked their location in the moving truck and was ready to move.

The squad silently arose and began to walk. The night was cold and all were glad to be moving. The Marines stopped frequently. These pauses allowed them to listen for the enemy; it was difficult to listen as effectively while moving. The stops also allowed Hummel to check his land navigation. Trapp checked it, too. The squad moved, slowly and silently, for hours that night.

The Marines occupied their chosen ambush site several hours before dawn. They spent all day concealed, watching for a sign of the enemy, but they saw nothing. The squad did not remain more than one day in the same place. They were careful to leave no trace of their presence either in their hide sites or in their ambush positions.

Each Marine had started the mission with 4 quarts of water and some food. As the mission went on, the Marines' packs got lighter and their belts got tighter. The water ran out after several days. Everyone shared what they had until each Marine was down to less than a quart, then the squad moved to one of the few small streams that ran through the area. The stream was 10 kilometers away over rough terrain. It took most of one night just to get to the stream and fill up the squad's canteens. The water tasted bad with the iodine necessary to purify it, but every water receptacle was full. Although they planned on topping off their canteens the next night before moving out, no one drank the water too greedily; they had to make it last in the event that they were unable to fill up again tomorrow. They all knew the allowance was one-and-a-half quarts per man per day, away from water sources. The group occupied an ambush site near the water that night, but saw nothing but a few small animals that came to drink. They now knew where they could find meat if they needed food.

In the evening on the eighth day of the patrol, a signal came down the tug line, a long piece of string that literally linked every Marine in the ambush site: someone was entering the kill zone from the northwest. From his position near the center of the ambush site, Trapp had no idea if it was the enemy or not or how many there were. He could only wait and see. He heard them before he could see them. They were speaking in hushed tones. He could not understand their language, but certain of their invulnerability, they were making little effort to be silent in the deepening twilight. There were six men. They were armed with a mixture of AK-47s and RPGs; these men were not shepherds out looking for a lost lamb.

The men were close. It seemed to Trapp that he could reach out and touch them, but he reassured himself that they were still at least

30 meters away. They passed in front of him from right to left headed toward an area that Trapp and Hummel both agreed would be an excellent position from which to ambush coalition convoys through the area.

Trapp was concerned that the men were going to get away. He was about to initiate the ambush himself when the world seemed to explode around him. The roar was deafening, but it ended as quickly as it had begun. Around him, Trapp could hear his Marines moving quickly. Two Marines swept through the kill zone, checking the bodies for anything of intelligence value. A few minutes later, one of the Marines came over and crouched beside him and Hummel.

"We got all six. It was hard to miss them, they were so close! Five are dead, but the sixth is alive. I've got Doc working on him right now. He's in rough shape, but if we get him back to the FOB, Doc thinks we can save him. We've secured all the weapons and we found some papers."

Hummel took all this in and rapped out his orders in a firm, but quiet voice. "Good. Make sure to get DNA samples. We'll move in two minutes." Turning to Trapp, Hummel continued, "Sir, I recommend we call in a MEDEVAC bird and get this guy out. We can put one of the Marines on it with him to take all of the intel back and do an initial debrief with the Intelligence Officer. We can make our water last at least another 2 days. If we stay out, we might catch some of the bad guys coming to look for their friends."

Trapp had been thinking the same thing. Hearing these words from Hummel reinforced Trapp's high opinion of the young Marine.

"I agree," said Trapp. "Where do you think we ought to go next?"

"Well, sir, I think we ought to stay nearby and let them come to us. They won't expect us not to move, and they won't be able to see us here even if they're looking."

When Trapp and the squad got off the helicopter at the FOB several days later, it looked like a mob had assembled to greet them. The noise of the helicopter drowned out every other sound. Lieutenant Colonel von Stahremberg stepped away from the group and advanced toward Trapp. In the dark, Trapp could not see that von Stahremberg was smiling until he was quite close.

"Congratulations, Trapp! Great job out there. Your Marines really came through!" exclaimed von Stahremberg.

"Thanks, sir. In addition to that ambush a couple days ago, we ambushed another group of enemy yesterday. There were about 20 of them. We killed eight that we know of; they left the bodies behind. We found a couple items we want the Intelligence Officer to take a look at. The rest ran as soon as we opened up, but there were a bunch of blood trails." If his Marines' aim had been a little better, there would have been no enemy survivors. Trapp made a mental note to adjust his training program.

"Your Marines managed to kill more of the enemy in two weeks than the rest of the battalion has in the previous two months, and all without a single casualty or any civilians hurt!"

Trapp was quickly jerked back to reality. His squad had not radioed back to the battalion, except to set up the MEDEVAC and the final exfiltration. Those calls had been extremely brief. He had little idea how the other squads had fared. As he now found out, each had laid at least one successful ambush. Better, none of his Marines had been injured. His feeling of pride in his men and relief did not last long, however. He realized that somewhere out in the darkness, the enemy was already working to find a way to counter his tactics and to get revenge for the Marines' recent successes.

"Sir, remember that the enemy learns. I need to get to work training my Marines to operate as part of a platoon in case the enemy tries to mass their forces. This will allow us to do a lot of other ambushes. I once read about X and Z shaped ambushes…"

The battalion commander put a hand on his shoulder and brought him up short.

"Whoa, whoa, whoa! I absolutely agree with you, but there's one other thing we have to do, too. I want to convert the rest of the battalion to true light infantry as well. Get with the Operations Officer as soon as you can and help him come up with a plan to retrain the other companies."

Chapter 7

Training Light Infantry Units

This chapter assumes you are the commander of a line infantry unit—platoon, company, or battalion—that you want to convert to light infantry. How would you do it? The same way Captain Trapp did.

Flexibility

The first step must be to give your troops a light infantry mindset. The way in which light infantry think is much different from the line infantry mindset. Retraining your men without changing the way that they think will give you light infantry in name only.

Changing the mindset of your men is not a "one-off" event. It must start immediately and continue throughout training. One part of this is an ongoing education program to teach troops about the basics of light infantry. Such an education program may consist of guided professional reading with linked discussions, tactical decision games, sand-table exercises, and tactical exercises without troops.

Another important method is to create situations that compel leaders to adapt to unexpected and constantly changing circumstances. Such situations should arise randomly, not just during scheduled training. Change the training schedule during the training. When units are on a mission during a field exercise, radio them and change their situation or mission and see how well they adapt. The new mission should be one for which they did not prepare and have little or no specialized gear. Run them through problems in the Field Leaders' Reaction Course (FLRC), if your station has one. The best book on how to train for adaptability is Don Vandergriff's *Raising the Bar*. Vandergriff spells out in detail why adaptability is so critical and how to inculcate it in your subordinates.

Some of your troops will thrive in this environment. Others will not. They joined the military for the order and structure they thought it would provide them. To reduce uncertainty, they will seek sources of "gouge." Be careful entrusting anyone with information about upcoming events, especially company clerks! Your real training plan should exist only in your own (paper, not electronic) notebook.

Free-Play Training

The next step that you, the commander, take is to make virtually all training free-play. The best way to train your unit is to replicate the conditions of combat as closely as possible. The best method for doing so is free-play training. One of the salient features of war is that it is a clash of opposing wills. Training that does not incorporate this will not be effective in preparing units for combat. On the rare occasions that troops get the opportunity to act freely as "aggressors" during current training exercises, they unleash their creativity and often cause great difficulty for their opponents. The philosophical goal for training light infantry is to make this "aggressor" mindset the mindset of your men all the time.

Weapons Proficiency

Third, ensure your troops are proficient with every weapon that they are likely to use in combat, including enemy and improvised weapons. Shoot under conditions that approximate combat (e.g., unknown distance at night) and evaluate how well each man shoots under all of these conditions. Do not succumb to the trap of conducting "familiarization" shoots. They are a waste of ammunition. Teach men how to disassemble, clean, reassemble, conduct a function check, and take immediate and remedial action for every weapon. Teach them how to inspect ammunition for serviceability, particularly ammunition for threat weapons. There should also be at least one designated marksman (DM) per squad. Even one well-trained DM can have a devastating effect upon the enemy. The DMs should be appropriately trained and provided with sniper rifles.

Every light infantryman should also be well-trained and comfortable using hand grenades. Ambushes are the preferred offensive and defensive method for light infantry and light infantry almost always operates in close terrain. Most combats are likely to be at short range. In such fighting, grenades prove extremely useful. Light infantrymen who are unfamiliar with or uncomfortable employing hand grenades will be at a disadvantage during such fighting.

Learning to Operate Patiently

Next, teach your men patience. Because light infantry is primarily foot-mobile and must remain concealed, even while moving, it will take time to obtain results. This change in the pace of operations must be reflected in the way units are trained. Troops should not be told when a field exercise is to end, nor should the training plan schedule an "ENDEX." Units should go to the field on the understanding that once there, they will be required to remain and to sustain themselves until they complete their mission.

To inculcate patience in your troops, avoid issuing orders that specify a time when something must be accomplished. Allow the unit leader to determine his own timeline. This timeline should be driven by tactical considerations, such as the time it takes to move stealthily or to conduct covert surveillance of an objective. The timeline should never be driven by the fact that an exercise must accomplish 13 training objectives in four days.

Stealth and Stalking

Many enlisted infantrymen hunt. Your training should build on the ways they know to hunt. Operating patiently and hunting skills, including stealth and stalking, go hand-in-glove. Light infantry that does not hunt its enemies because it has poor stalking skills are more likely to get ambushed than to ambush. Operating away from the aid of other friendly units, light infantry must rely on superb field skills to survive. To be observed is to invite attack and destruction. A question to ask your men frequently in training is, "Will what you are doing here make you the hunters or the hunted?"

The best way to train troops in stealth and stalking is to let the experienced hunters lead and critique the others, regardless of rank. It is likely that there will be several who possess superior fieldcraft. As always, the best way to build these skills is for men to take part in free-play force-on-force exercises. Your troops' competitive nature will be unleashed and each unit will strive to hunt better than the others because those who have the best stealth and stalking skills will usually win.

Survival Training

Light infantry will often be forced to live off the land. Sometimes, this will mean buying food from local merchants and farmers. But the ability to identify plants and animals that will sustain life should be taught, and these skills should be regularly exercised. Troops should be taught how to purify water from streams and lakes. Light infantry does not depend on bottled water. During exercises, units should not be regularly resupplied. This will increase their ability to live off the land and force them to make do with the items on hand.

Physical Fitness

Light infantrymen require a level of physical fitness that is both greater than and different from their line infantry counterparts. Light infantrymen must be able to march great distances rapidly while carrying mission-essential gear. The minimum sustained march rate is 40 kilometers per day; historically, some light infantry units have sustained rates as high as 70–80 km daily. Physical fitness events that build such performance should be incorporated into each training exercise. Running is irrelevant and a waste of time.

In order for light infantry to be mobile, the gear load must be strictly maintained at no more than 45 pounds. Studies over centuries have shown that weights greater than 45 pounds (total including individual clothing, weapons, and other equipment) rapidly degrades an individual's ability to march great distances. Not only must light infantrymen be prepared to make long foot movements, they must be prepared to fight once they arrive. There can be no individual exceptions to the weight limit; the unit's march performance will be that of its slowest member. However, where circumstances permit, light infantry units can and do make use of carts, bicycles, and pack animals to carry heavier loads.

Physical fitness, like the light infantry mindset, is an ongoing training goal. Not only can physical training be a stand-alone event, it should also be part of every activity. Units should march most of the places they go.

Demolitions

Train your troops to use demolitions, to the point where they are both comfortable and creative with them. Demolitions are of inestimable use in initiating an ambush and can also be used to destroy enemy equipment following the ambush if it cannot be carried away. During an attack, demolitions can be used to breach enemy obstacles to permit the assault force to penetrate the defense. You should use IEDs better than your enemy uses them against you. The small size of light infantry units and the need to conduct demolitions quickly makes it imperative that every man is trained to conduct them.

Land Navigation

Land navigation is a critical skill for all hunters. Land navigation practice, both day and night, should begin very early in the training program. Each training exercise should consider how to incorporate missions that will challenge and develop land navigation skills in the unit. Unit leaders should ensure that navigation duties are rotated throughout the unit.

The issue is not just technical skills. Light infantry need the ability to know instinctively where they are. Electronic land navigation aids work against this. You should train without them. Electronic aids also require batteries. Stocks of batteries add weight and take up space in troops' packs, not to mention requiring resupply missions that could compromise the unit's position.

Light infantrymen must become expert in land navigation using a map and compass. No electronic aids of any kind should be permitted. Every individual must be made to demonstrate his ability to navigate effectively. Do not allow any of the troops to "hide" and rely on their comrades. Their life or the lives of their friends may one day depend on how well they navigate, particularly at night, which is when light infantry often moves. No light infantryman who lacks an instinctive sense for his location should serve in a leadership position.

Surveillance / Tactical Site Exploitation

Light infantry units must be experts in surveillance and must be able to discern the slightest weaknesses in the enemy's positions or posture. Light infantry tactics rely on exploiting such weaknesses. Troops must also be able to communicate this information clearly and succinctly to others. The ability to draw a quick sketch of the situation is valuable and should be developed in training.

Men must also be trained to pick up items of intelligence value following successful attacks or ambushes. Troops must know what to take, how to preserve it, and how to catalog it for later exploitation. They should also be trained to cover the fact that they have found and taken material with intelligence value. Intel the enemy does not know we have is the best kind.

Medical Training

Light infantry units should be taught to treat and care for casualties until they can be evacuated. In some circumstances, evacuation may take several days. Medical training, once taught, should be incorporated into every field exercise. Troops should be forced to treat, transport, and evacuate casualties until doing so is second nature.

Supporting Arms

The lowest training priority should be given to teaching your men to utilize supporting arms. This is not because supporting arms are not useful. While light infantry relies on its own weapons, it does make use of supporting arms when they are available. Light infantrymen should be trained to call for and adjust indirect fires and to communicate with close support aircraft, but the way they do these tasks are essentially the same as for line infantry.[8] Some of your troops, certainly your junior officers and staff noncommissioned officers, should have received this training in school. They can teach others.

Conclusion

In training light infantry units, it is critical to consider how these units will be employed. In Fourth Generation war, light infantry generally operates as small units. It is therefore important to focus light infantry training at the lowest levels first and work upward. Good teams contribute to good squads. Good squads contribute to good platoons, and so on. But most operations against non-state forces will be at the platoon level or lower.

Chapter 8

Light Infantry Conversion Training Plan

You have decided to retrain your line infantry unit as light infantry for Fourth Generation war. How do you go about it? Here is a proposed training plan. The proposed training plan makes the following assumptions:

1. Each soldier or Marine has completed basic military occupational specialty (MOS) training (although some of this training may have been counter-productive from a light infantry perspective). This means that troops know basic battlefield first aid, have achieved basic proficiency with their individual weapon, and so forth.

2. This training package is designed to be conducted with a minimum of external augmentation (other than in areas requiring subject matter expertise not usually resident at the company level). It presumes little understanding of the light infantry concept on the part of the unit's leaders and enlisted personnel. Due to this knowledge gap and the different role of officers

and senior noncommissioned officers, at some point the company commander must consider a separate education track for more senior personnel. This could occur from the very beginning of the training program. At the very least, this dual track should begin at the start of the platoon training phase to prepare lieutenants and senior enlisted personnel for the roles they will play in light infantry units operating above the squad level.

3. Light infantry training must operate within existing Table of Equipment (T/E) constraints. While converting from line infantry to light infantry will require some changes in gear, it is assumed that there is no money for large, expensive programs.

4. Not all training events are captured in the following training schedule. No attempt is made to create a detailed hour-by-hour plan. Only major training events are captured each day. For instance, physical training events (other than programmed hikes) are not shown on the schedule.

5. As much as possible, classes should be conducted in the field with practical application immediately following.

Week	Schwerpunkt	Monday	Tuesday	Wednesday	Thursday	Friday	Saturday	Sunday
1	Phase I: Individual Skills **Introduction to LI**	Basics of LI LI TDG 1	Field Leader's Reaction Course (FLRC)	Land Navigation I (classroom)	Land Navigation I (practical application)	Surveillance I Basic Survival Skills I		Church services Possible surprise training event
2	Phase I: Individual Skills **Survival and Tracking**	LI TDG 2 Basic Tracking I	Basic Survival Skills II (classes and practical application)	Basic Tracking II	Basic Survival Skills III (classes and practical application)	6 mile conditioning hike	Training remediation time	Church services Possible surprise training event
3	Phase I: Individual Skills **Medical Training**	Medical Training (classes and practical application)	Medical Training (classes and practical application)	Medical Training (classes and practical application)	Medical Training (classes and practical application)	Medical Training (review and evaluation)	Possible surprise training event	Church services Possible surprise training event
4	Phase II: Weapon Skills **Field Firing Techniques**	Communications (classes and practical application)	Hike to unknown distance range (day / night fire) Advanced Marksmanship classes	Day / night fire	Unknown distance firing, use of optics, NVDs	9 mile hike back from range Weapons Cleaning	Possible surprise training event	Church services Possible surprise training event
5	Phase II: Weapon Skills **Threat Weapons**	Advanced Survival Skills I LI TDG 3	Hike to range Threat Weapons classes at range Threat Weapons live fire (day / night)	Threat Weapons live fire (day / night)	Threat Weapons Live Fire	Hike back from range Weapons Cleaning	Possible surprise training event	Church services Possible surprise training event

Week	Schwerpunkt	Monday	Tuesday	Wednesday	Thursday	Friday	Saturday	Sunday
6	Phase II: Weapon Skills **Hand Grenades**	Land Navigation II (classroom)	Land Navigation II (practical application)	Hike to grenade range Throw practice, live grenades	Hike to fortified position range Throw practice, live grenades in fortified positions	12 mile ike back from hand grenade range	Possible surprise training event	Church services Possible surprise training event
7	Phase II: Weapon Skills **Demolitions**	LI TDG 4 Basic Demolitions	Hike to demolition range Conduct demolitions	Demolition range	Demolition range	Hike back from demolition range	Possible surprise training event	Church services Possible surprise training event
8	Phase III: Unit Skills **Ambush Fundamentals**	Light Infantry TDG 5 Ambush Sand Table Exercise	Land Navigation III (classroom)	Land Navigation III (practical application)	Training Remediation	Field Preparation	Training skills remediation	Church services
	Senior Leaders			*Ambush Tactical Exercise Without Troops*				
9	Phase III: Unit Skills **Field Exercise**	Exercise Preparation	Field Exercise I Hike to laager site	Field Exercise I Receive mission in laager site	Field Exercise I Conduct reconnaissance	Field Exercise I Conduct mission	Field Exercise I Receive mission in laager site	Field Exercise I Conduct reconnaissance
10	Phase III: Unit Skills **Field Exercise**	Field Exercise I Conduct mission	Field Exercise I Receive mission in laager site	Field Exercise I Conduct reconnaissance	Field Exercise I Conduct mission	Hike back from training area Weapons maintenance		Church services Possible surprise training event

Week	Schwerpunkt	Monday	Tuesday	Wednesday	Thursday	Friday	Saturday	Sunday
11	Phase IV: Weapon Skills II **Crew Served Weapons Employment** *Senior Leaders*	LI TDG 6 Machine gun employment	Land Navigation IV *MG/Supporting Arms Employment for Light Infantry Discussion*	Call for Fire Light mortar employment	Call for Fire Prac. App. Supporting Arms Employment Sand Table Exercise	Aviation Integration	Possible surprise training event	Church services Possible surprise training event
12	Phase IV: Weapon Skills **Crew Served Weapons Live Fire**	Hike to mortar range Call for fire/light mortar live fire and employment	Call for fire/light mortar live fire and employment	Light mortar live fire Hike to machine gun range	Machine gun live fire / live fire night ambush with machine guns	Machine gun live fire Hike back from machine gun range		Church services
13	Phase IV: Weapon Skills **Light Infantry Squad Live Fire**	Hike to training area Squad ambush field exercise	Squad ambush field exercise	Squad ambush field exercise	Squad ambush field exercise Hike to live fire range	Squad ambush live fire (day/night)	Squad ambush live fire (day) Hike back from range	Church services Possible surprise training event
14	Phase V: Platoon Phase **Light Infantry Platoon** *Senior Leaders*	LI TDG 7 Medical sustainment training *Platoon Offense STEX*	Land Navigation V *Platoon Offense TEWT*	Squad competition *Platoon Defense STEX*	Squad competition *Platoon Defense TEWT*	Field preparation time	Training skills remediation	Church services
15	Phase V: Platoon Phase **Field Exercise**	Hike to training area	Platoon live fire attacks	Platoon live fire ambush (day/night)	15 mile hike back from training area	Possible surprise training event	Training skills remediation Possible surprise training event	Church services Possible surprise training event

Week	Schwerpunkt	Monday	Tuesday	Wednesday	Thursday	Friday	Saturday	Sunday
16	Phase V: Platoon Phase **Light Infantry Platoon**	Field Preparation	Platoon Field Exercise	Platoon Field Exercise	Platoon Field Exercise	Platoon Field Exercise	Platoon Field Exercise	Platoon Field Exercise (Including field church service)
17	Phase V: Platoon Phase **Field Exercise**	Platoon Field Exercise	Platoon Field Exercise	15 mile hike back from training area	Weapons/gear maintenance	Possible surprise training event	Training skills remediation / Possible surprise training event	Church services / Possible surprise training event
18	Phase VI: Company Phase **Light Infantry Company**	LI TDG 8 / Supporting arms sustainment training	Intelligence training sustainment	Squad competition in weapons skills	Squad competition in medical / survival skills	LI TDG 9	Possible surprise training event	Church services
	Senior Leaders	*Company Offense Sand Table Exercise*	*Company Offense Tactical Exercise Without Troops*	*Company Defense Sand Table Exercise*	*Company Defense Tactical Exercise Without Troops*			
19	Phase VI: Company Phase **Field Exercise**	Company Field Exercise	Company Field Exercise	Company Field Exercise	Company Field Exercise	Company Field Exercise	Company Field Exercise	Company Field Exercise
20	Phase VI: Company Phase **Field Exercise**	Company Field Exercise	Company Field Exercise	Company Field Exercise	15 mile hike back from training area			Church services

Notes:

- The FLRC presents units with problems that require teamwork and resourcefulness to solve. They can be extremely difficult. This course gives the unit commander insight into the leadership, resourcefulness, imagination and problem-solving qualities that his junior leaders possess.

- Throughout this transition program, subordinates should be given incomplete information, and the training schedule should be changed at short notice. The purpose of such changes is to accustom the troops to rapid change and to engender in them an ability to adapt.

- The threat weapons package consists of assembly, disassembly, maintenance, immediate, and remedial action on AK-47, PKM, and RPG (at a minimum).

- Field Exercises should be permitted to last longer than programmed in the above schedule, based upon the pace of the unit in conducting missions (although this may impact follow-on training). In order to force leaders to become accustomed to rapid changes, at least twice during the exercise units should be re-tasked after they have completed mission planning. Students should be given no water beyond their initial load and only 4 days of food. Units should be forced to exercise their survival training and get used to living off the land while conducting operations.

Chapter 9

Meeting the Fourth Generation War Challenge

Just as Fourth Generation war represents the biggest change in warfare since the Peace of Westphalia in 1648, it also represents the greatest challenge. What can the armed forces of the state do to meet that challenge?

First, if they are to have any hope of meeting it, they must be or become Third Generation, maneuver warfare militaries. At present, few if any are. Becoming a Third Generation service is not simply a matter of adopting infiltration tactics. At root, it requires a change of institutional culture.

First and Second Generation militaries worship the culture of order. They focus inward on orders, regulations, processes, and procedures; decision-making is centralized; they prize obedience over initiative; and they rely on imposed discipline. Third (and most non-state Fourth) Generation militaries know that the culture of order, which originated on the orderly battlefields of the 18th century, is outdated. They focus outward on the situation, the enemy, and the result the situation demands; decision-making is decentralized; they prize initiative over obedience; and they depend on self, not imposed, discipline. Commanders are held responsible for results but never for method.

Making the cultural change from the Second to the Third Generation is very difficult. Defenders of the culture of order are numerous and tend to become more so as rank increases. Militaries draw such people, and a challenge to the culture of order is a threat to what their psychology demands.

However, a Second Generation military has no chance of victory in a Fourth Generation war. Any armed service that fails to transition to the Third Generation is doomed to irrelevance as Fourth Generation war spreads. While it is not clear how successful even Third Generation state-armed forces will be at Fourth Generation war, it is clear that Second Generation militaries will fail. Fourth Generation war cannot be reduced to procedures for putting firepower on targets.

If an armed service is given the mission of intervening in a Fourth Generation war in another country with the objective of saving or restoring a state, it is almost certain to lose. This is true even if it does everything right. The strategic factors working against it will almost always be too powerful to overcome. It represents a foreign country, a different culture, often a different religion. It does not know the local culture well, it does not understand local politics, its troops do not speak the language. If it comes from a First World country, it represents the rich destroying what little the poor have. At some point it will go home, leaving the locals who worked with it open to retaliation as collaborators. Most powerfully, its very presence undermines the legitimacy of the state it is attempting to support. Because legitimacy is the ground on which Fourth Generation war is fought, any foreign intervention force is undermining the local state more powerfully than it can hope to buttress it. A mission to uphold or restore a failing or failed state is, for any armed service coming from outside that state, a poisoned chalice.

The mission for which the armed forces of the state must prepare, and in which they must succeed or perish, is defending and preserving the state at home. While the crisis of legitimacy of the state varies greatly in intensity, it now affects almost all states to some degree. It is likely to become more intense over time, in First World countries as well as the Third World countries where it is presently observed. Many First World countries, in an act of folly almost without precedent, have imported Fourth Generation war by the literal shipload as they admitted millions of immigrants and refugees from other cultures. Some of those immigrants and refugees will refuse to acculturate, often on religious grounds. Others might be willing to do so, but are arriving in numbers so great they overwhelm the acculturation process. These immigrants offer a base for Fourth Generation war on the soil of any country that receives them.

Other, domestic developments also point toward Fourth Generation war within a growing number of states. As citizens transfer their primary loyalty away from the state, two recipients of that loyalty are also likely bases for Fourth Generation war. The first is gangs, which are becoming more powerful all over the world. Many are successful if illegal economic enterprises, which means they have the money for war. In an increasing number of Third World countries, the state is no longer sufficiently powerful to defeat the gangs. Rather, it must make such deals with them as it can get.

A second domestic source of Fourth Generation war is "causes," strong emotional attachments to alternative loyalties that range from religious sects and ideologies to "animal rights." While few of these have the resources at gangs' disposal, they can engender fanatic loyalty. Not many gang members are likely to become suicide bombers (it seldom pays), but people deeply attached to a "cause" may do so. Like gangs, "causes" battle not only the state but each other, creating disorder that further undermines the legitimacy of the state.

Fourth Generation war on a state military's home soil offers a challenge it must meet. While overseas interventions in Fourth Generation war are usually cabinet wars, the loss of which means little, Fourth Generation war at home poses an existential threat. If a state's armed services cannot defeat that threat, the state will disappear and its armed forces with it. In their place will come chaos. As was the case in Europe between the end of the Middle Ages and the rise of the state, life will be nasty, brutish, and short.

Fortunately for the state and its armed forces, Fourth Generation war at home is significantly easier to win than Fourth Generation war abroad. The strategic and moral factors that work so strongly against a state military overseas diminish or vanish. At home, the state military represents the home country, its own culture and its religion. The troops are often local young men. It knows the culture and language. It is not leaving, because it is already home. Its presence, if it acts as this handbook recommends, can bolster rather than undermine the state's legitimacy.

However, while there is a great deal a state military can do to preserve its own state, there is one decisively important thing it cannot do. It cannot provide competent governance. The main organs of the state's government and the civilians who head and run them must be competent. They must do what states exist to do, above all providing order: safety of persons and property. They must make things work: the police, the courts, the schools, the country's infrastructure, and increasingly its economy as well (having accepted major economic responsibilities, the legitimacy of a state now depends in part on how well it manages the economy). Corruption cannot be so far-reaching as to destroy the state's ability to do its duties, nor to the point where it is a public scandal. Whether a state is "democratic" or not matters little. Most people accept as best the government that governs best. A government that fails to govern effectively will cost the state legitimacy, regardless of whether it is democratic or autocratic.

As state militaries give over the jousting contests that wars between states have mostly become and reorient to confront Fourth Generation war on their own soil, they will find themselves dependent on a condition they cannot create: good governance. That may prove the most difficult challenge of all.

Appendix A: The First Three Generations of Modern War

The Chinese military philosopher Sun Tzu said, "He who understands himself and understands his enemy will prevail in one hundred battles." In order to understand both ourselves and our enemies in Fourth Generation conflicts, it is helpful to use the full framework of the Four Generations of modern war. What are the first three generations?

First Generation war was fought with line and column tactics. It lasted from the Peace of Westphalia until around the time of the American Civil War. Its importance for us today is that the First Generation battlefield was usually a battlefield of order, and the battlefield of order created a culture of order in state militaries. Most of the things that define the difference between "military" and "civilian"—saluting, uniforms, careful gradations of rank, etc.—are products of the First Generation and exist to reinforce a military culture of order. Just as most state militaries are still designed to fight other state militaries, so they also continue to embody the First Generation culture of order.

The problem is that, starting around the middle of the 19th century, the order of the battlefield began to break down. In the face of mass armies, nationalism that made soldiers want to fight, and technological developments such as the rifled musket, the breechloader, barbed wire, and machine guns, the old line-and-column tactics became suicidal. But as the battlefield became more and more disorderly, state militaries remained locked into a culture of order. The military culture that in the First Generation had been consistent with the battlefield became increasingly contradictory to it. That contradiction is one of the reasons state militaries have so much difficulty in Fourth Generation war, where not only is the battlefield disordered, so is the entire society in which the conflict is taking place.

Second Generation war was developed by the French Army during and after World War I. It dealt with the increasing disorder of the battlefield by attempting to impose order on it. Second Generation war, also sometimes called firepower/attrition warfare, relied on centrally controlled indirect artillery fire, carefully synchronized with infantry, cavalry and aviation, to destroy the enemy by killing his soldiers and blowing up his equipment. The French summarized Second Generation war with the phrase, "The artillery conquers, the infantry occupies."

Second Generation war also preserved the military culture of order. Second Generation militaries focus inward on orders, rules, processes, and procedures. There is a "school solution" for every problem. Battles are fought methodically, so prescribed methods drive training and education, where the goal is perfection of detail in execution. The Second Generation military culture, like the First, values obedience over initiative (initiative is feared because it disrupts synchronization) and relies on imposed discipline.

The United States Army and the U.S. Marine Corps both learned

Second Generation war from the French Army during the First World War, and it largely remains the "American way of war" today.

Third Generation war, also called maneuver warfare, was developed by the German Army during World War I. Third Generation war dealt with the disorderly battlefield not by trying to impose order on it but by adapting to disorder and taking advantage of it. Third Generation war relied less on firepower than on speed and tempo. It sought to present the enemy with unexpected and dangerous situations faster than he could cope with them, pulling him apart mentally as well as physically.

The German Army's new Third Generation infantry tactics were the first non-linear tactics. Instead of trying to hold a line in the defense, the object was to draw the enemy in, then cut him off, putting whole enemy units "in the bag." On the offensive, the German "storm-troop tactics" of 1918 flowed like water around enemy strong points, reaching deep into the enemy's rear area and also rolling his forward units up from the flanks and rear. These World War I infantry tactics, when used by armored and mechanized formations in World War II, became known as "Blitzkrieg."

Just as Third Generation war broke with linear tactics, it also broke with the First and Second Generation culture of order. Third Generation militaries focus outward on the situation, the enemy, and the result the situation requires. Leaders at every level are expected to get that result, regardless of orders. Military education is designed to develop military judgment, not teach processes or methods, and most training is force-on-force free play because only free play approximates the disorder of combat. Third Generation military culture also values initiative over obedience, tolerating mistakes so long as they do not result from timidity, and it relies on self-discipline rather than imposed discipline, because only self-discipline is compatible with initiative.

When Second and Third Generation war met in combat in the German campaign against France in 1940, the Second Generation French Army was defeated completely and quickly; the campaign was over in six weeks. Both armies had similar technology, and the French actually had more (and better) tanks. Ideas, not weapons, dictated the outcome.

Despite the fact that Third Generation war proved its decisive superiority more than 60 years ago, most of the world's state militaries remain Second Generation. The reason is cultural: they cannot make the break with the culture of order that the Third Generation requires. This is another reason why, around the world, state-armed forces are not doing well against non-state enemies. Second Generation militaries fight by putting firepower on targets, and Fourth Generation fighters are very good at making themselves untargetable. Virtually all Fourth Generation forces are free of the First Generation culture of order; they focus outward, they prize initiative and, because they are highly decentralized, they rely on self-discipline. Second Generation state forces are largely helpless against them.

Appendix B:
The 4GW Canon

There are seven books which, read in the order given, will take the reader from the First Generation through the Second, the Third and on into the Fourth. We call them the 4GW Canon.

The first book in the canon is C.E. White, *The Enlightened Soldier*. This book explains why you are reading all the other books. It is the story of Scharnhorst, the leader of the Prussian military reform movement of the early 1800s, as a military educator. With other young officers, Scharnhorst realized that if the Prussian army, which had changed little since the time of Frederick the Great, fought Napoleon, it would lose and lose badly. Instead of just waiting for it to happen, he put together a group of officers who thought as he did, the *Militaerische Gesellschaft*, and they worked out a program of reforms for the Prussian state and army. Prussia's defeat at the battle of Jena opened the door to these reforms, which in turn laid the basis for the German army's development of Third Generation war in the early 20th century.

The next book is Robert Doughty, the former head of the History Department at West Point and the best American historian of the modern French Army. *The Seeds of Disaster* is the definitive history of the development of Second Generation warfare in the French army during and after World War I. This book is in the canon because the U.S. Army and Marine Corps learned modern war from the French, absorbing Second Generation war wholesale. As late as 1930, when the U.S. Army wanted a manual on operational art, it took the French manual on Grand Tactics, translated it and issued it as its own. *The Seeds of Disaster* is the only book in the canon that is something of a dull read, but it is essential to understanding why the American Armed Forces act as they do.

The third book, Bruce Gudmundsson's *Stormtroop Tactics*, is the story of the development of Third Generation war in the German Army in World War I. It is also a book on how to change an army. Twice during World War I, the Germans pulled their army out of the Western Front unit-by-unit and retrained it in radically new tactics. Those new tactics broke the deadlock of the trenches, even if Germany had to wait for the development of the Panzer divisions to turn tactical success into operational victory.

Book four, Martin Samuels's *Command or Control?*, compares British and German tactical development from the late 19th century through World War I. Its value is in the clear distinctions it draws between the Second and Third Generations, distinctions the reader will find useful when looking at the U.S. Armed Forces today. The British were so firmly attached to the Second Generation—at times, even the First—that German officers who had served on both fronts in World War I often said British troop handling was even worse than Russian. Bruce Gudmundsson says that in each generation, only one Englishman is allowed to truly understand the Germans. In our generation, Martin Samuels is that Brit.

The fifth book in the canon is another one by Robert Doughty: *The Breaking Point*. This is the story of the battle of Sedan in 1940, where Guderian's Panzers crossed the Meuse and then turned and headed for the English Channel in a brilliant example of operational art. Here, the reader sees the Second and Third Generations clash head-on. Why does the Third Generation prevail? Because over and over, at decisive moments, the Third Generation Wehrmacht takes initiative (often led by NCOs in doing so) while the French wait for orders. What the French did was often right, but it was always too late.

The sixth book in the canon is Martin van Creveld's *Fighting Power*. While *The Breaking Point* contrasts the Second and Third Generations in combat, *Fighting Power* compares them as institutions. It does so by contrasting the U.S. Army in World War II with the German Army. What emerges is a picture of two radically different institutions, each consistent with its doctrine. This book is important because it illustrates why you cannot combine Third Generation, maneuver warfare doctrine with a Second Generation, inward-focused, process-ridden, centralized institution.

The seventh and final book in the canon is Martin van Creveld's, *The Transformation of War*. Easily the most important book on war written in the last quarter-century, *Transformation* lays out the basis of Fourth Generation war, the state's loss of its monopoly on war and on social organization. In the 21st century, as in all centuries prior to the rise of the nation-state, many different entities will fight war, for many different reasons, not just *raison d'etat*. Clausewitz's "trinity" of people, government, and army vanishes, as the elements disappear or become indistinguishable from one another. Van Creveld has also written another book, *The Rise and Decline of the State*, which lays out the historical basis of the theory described in *Transformation*.

Appendix C: Light Infantry Essential Reading

A well-rounded educational plan for light infantry leaders should incorporate as many of the following works as possible:

- Austro-Hungarian Marine Corps. *Fleet Marine Force Manual 2 (FMFM-2): Light Infantry*. Imperial and Royal Publishers: Vienna, 2008. Available at http://www.traditionalright.com/resources.

- Canby, Steven L. *Classic Light Infantry and New Technology*. C&L Associates Report, n. p., 1983. (DOD Contract No. MDA 903-81-C-0207)

- Ewald, Johann von. *Diary of the American War: A Hessian Journal*. Ed. by Joseph P. Tustin. New Haven: Yale University Press, 1979.

- Ewald, Johann von. *Treatise on Partisan Warfare*. Trans. by Robert A. Selig and David Curtis Skaggs. Contributions in Military Studies, Number 16. New York: Greenwood Press, 1991.

- Fuller, J. F. C. *British Light Infantry in the Eighteenth Century.*

- Lind, William S. "Light Infantry Tactics." *Marine Corps Gazette* (June 1990), 42–.

- McMichael, Scott R. *A Historical Perspective on Light Infantry.* Combat Studies Institute Research Survey No. 6. Washington, D.C.: Government Printing Office, 1987. A PDF of this monograph may be found at: `http://usacac.army.mil/cac2/cgsc/carl/download/csipubs/HistoricalPerspectiveonLight Infantry.pdf`

- Schmitt, John F. "Light Infantry Tactics at the Company Level and Below." *Marine Corps Gazette* (June 1990), 48–.

- Vandergriff, Donald E. *Raising the Bar: Creating and Nurturing Adaptability to Deal with the Changing Face of War.* Washington, D.C.: Center for Defense Information, 2006.

- Uhle-Wettler, Franz. *Battlefield Central Europe: The Danger of Overreliance on Technology by the Armed Forces,* 1981. This book can still be found for sale on the internet.

Notes

1. For a description of the first three generations, see Appendix A.

2. Republished with two follow-up pieces in the November, 2001 *Marine Corps Gazette*.

3. Martin van Creveld, *The Rise and Decline of the State* (Cambridge University Press, Cambridge, U.K.; 1999).

4. The Israeli military historian Martin van Creveld calls this kind of war "non-trinitarian warfare," because it does not fit within Clausewitz's trinity of government, army, and people where each of those elements is related but distinct.

5. See Col. Thomas Hammes, *The Sling and the Stone*.

6. Conversation between Martin van Creveld and William S. Lind, May 2004, Bergen, Norway.

7. Colonel David H. Hackworth, About Face (Simon and Schuster, New York, 1989) pp. 679–680.

8. In order to support true light infantry most effectively, pilot training and even aircraft types must change. Back in the mid-1990s, the U. S. Navy and Marine Corps experimented with a

concept called "Jaeger Air." The objective was to provide aviation support to units in a maneuver warfare (Third Generation war) environment. While this experiment was cut short, it showed great promise and should be revived. See also the K.u.K. Air Cooperation Field Manual FMFM 3-23, available at https://www.traditionalright.com/resources.